Disaster and Gender in Coastal Bangladesh

Mohammed Moniruzzaman Khan

Disaster and Gender in Coastal Bangladesh

Women's Changing Roles, Risk and Vulnerability

Springer

Mohammed Moniruzzaman Khan
Department of Sociology
Jagannath University
Dhaka, Bangladesh

ISBN 978-981-19-3283-0 ISBN 978-981-19-3284-7 (eBook)
https://doi.org/10.1007/978-981-19-3284-7

This Springer imprint is published by the registered company Springer Nature Singapore Pte Ltd.
The registered company address is: 152 Beach Road, #21-01/04 Gateway East, Singapore 189721, Singapore

This book is dedicated to the loving memory of my father Abdur Rahman Khan and my mother Ayesha Rahman who sacrificed everything for our happiness and smiling faces.

Preface

This book comprises seven chapters; Chapter 1 of the book focuses on the research background by introducing the importance of the study of disaster, women and vulnerability discourse in coastal Bangladesh. It explains the background, rationale and aims of the study and focuses on the socio-economic and demographic context of women in Bangladesh; history of disasters in Bangladesh; disaster, vulnerability, coping and adaptation mechanisms by women; and institutionalisation of disaster management, women and climate change policies. Moreover, this chapter attempts to focus on women's role in disaster management policies, as well as the position of women as leaders in disaster management and how GO and NGOs popularise and strengthen women's institutional arena. Chapter 1 discusses the theories related to disaster, women, changes and vulnerability. Theoretical framework and conceptual considerations are discussed to understand how theoretical issues such as pressure and release, access and entitlement and women and social transformation are fit and relevant to research questions and objectives. This chapter basically works as a base that guides the complete study to focus on relevant theories and conceptualisations in the prevailing literature to identify the research gap and ways in which gender analysis in disaster research can be conceptualised. Chapter 2 discusses the sources of data, techniques, methods and plans for investigation. The chapter presents the background to the selection of research questions, study area and methodology for this study. The reasons for selecting Dacope, Khulna, as a suitable study area for the investigation are provided in Chap. 2. In addition to socio-economic conditions, this chapter focuses on the vulnerability to disasters in Dacope. In this chapter, the methodology and methods of data collection are elaborated. This chapter explains all the steps of data collection, analysis and challenges faced in performing fieldwork in the remote Dacope of Khulna. In this chapter, the manner in which challenges are dealt and consequent limitations of this research are discussed.

Chapter 3 focuses on the vulnerability of women in the coastal area in everyday life, particularly for social practices and traditions. Chapter 4 covers the vulnerability of women in the coastal area from the aspects of lacking access and entitlement to resources and not being able to play a decision-making role. Furthermore, this chapter deals with, along with coping and adaptation mechanisms, the different challenges and responses of women of different backgrounds. Chapter 5 focuses on women's empowerment, agency building and how the interventions of GO and NGOs are effective. However Chap. 6 addresses the status of women participation with the decision-making process at both community and family level, as well as their representation at different committees and their impacts. This chapter focuses on the extent to which the intervention of GO and NGO helps women build an agency and consequently women are able to play a more active role in the community. In the coastal area, although changes, social transformation, development of agency, their participation and involvement with decision-making process are not radical; however, it has happened in daily practices, traditions and everyday lives of women at a community level because of the intervention of GO, NGOs and *Aila* as a disaster. Finally, Chap. 7, the concluding chapter, links important empirical results relating to the theoretical explanation. Moreover, it offers recommendations on research implications and policy implications for speeding up and considering an inclusive and fair strategy for developing agency and empowering women. This chapter ends with certain suggestions for additional disaster-related studies, women and vulnerability and develops a model for understanding disaster, gender and improving/empowering coastal women.

Dhaka, Bangladesh Mohammed Moniruzzaman Khan

Declaration by the Author

This is my own work and acknowledges that all results, data and quotations have been used from the published or unpublished study of other people. There were no changes to intellectual content or substance of this book because of advice from third parties (paid or unpaid) who contributed by collecting data or provided copy editing or proof reading services. I completely acknowledged all such contributions. To my knowledge, this original work does not break any law or infringe any third party's copyright or intellectual property right.

Mohammed Moniruzzaman Khan

Acknowledgements

Bryant McGill metaphorically reported that nothing grows in the shadow of want without the sunlight of acknowledging one's fullness. From the front stage, I therefore have always looked forward to the opportunity of duly thanking each person who inspired me from the backstage in writing this book.

My foremost appreciation, deepest sense of gratitude and indebtedness to my Ph.D. supervisor Dr. Alison Loveridge for contributing to develop my intellectual engagement and style through her constructive criticism. Without her scholastic guidance, valuable suggestions, constant encouragement, constructive feedback and direct involvement with every chapter of the book whenever I asked for, it would have never been possible to publish this book. Professor Dr. Rangalal Sen, Professor Dr. Mijanur Rahman, Professor Dr. Mahbuba Nasreen, Professor Dr. Mokadem Hossain, Professor Dr. Mashiur Rahman, Dr. Debasish Kundu of Dhaka University taught me to examine sociology and social problems through the lens of others. I owe them for making me what I am today. I thank Jagannath University, Dhaka, for granting me leave for this research and bearing my absence from the academic point of view of its Department of Sociology, Faculty of Social Sciences.

I would like to thank all government officials in Bangladesh from national to local level, particularly UNO, Dacope, who helped me beyond my expectations to gain access in the research field. During my fieldwork, I have been extremely fortunate to have wonderful research assistants with whom I travelled all the time and who played an important role and worked really hard in my fieldwork in Bangladesh. My appreciation is extended to all respondents, participants and discussants who provided information and offered me friendship during field data collection. I sincerely acknowledge the cooperation and support of locally elected representatives, particularly the chairmen of both Sutarkhali and Kamarkhola unions.

Moreover, I would like to extend my sincere thanks to my elderly mother for her sacrifice; I am privileged to have had supportive and loving family members, which has greatly helped in shaping this study. Finally, there is one person who deserves my appreciation above all others and that is my wife Nurunnahar Mazumder who was kind, sacrificing, patient, loving and supporting. If she did not take all responsibilities of the family on her shoulders, none of this would have been possible. She deserves my sincerest acknowledgement for believing in me and for sacrificing, managing the family and for facing all types of problems in my absence.

Throughout this study, it would be less than fair if I did not acknowledge the support received from Dr. Farida Akhtar Khanam, Dr. Leema Hoque, Shipra Sarker, Dr. Farhana Zaman, Choyan Kumar Saha, Mostofa Mukul, Arif Rahman, Rifat Istiak, Shamima Nasreen Mily, Nazmul Islam Opu, Marjia Rahman Mili, Suborna Mondol, Suza Uddin, Sadman Sakib, Rahela Mukta, Mehraz Tapu, Md. Jahidul Islam, Esmin Sarkar Eva and Toma.

Contents

About the Editor

Professor Dr. Mohammed Moniruzzaman Khan is involved with teaching and research professionally since 1999 and serving currently as a core faculty member in the Department of Sociology, Jagannath University, 2nd largest public university in Bangladesh. He also serves as adjunct faculty in the Department of Political science and Sociology, North South University (NSU) and in the Department of Economics and Social sciences, BRAC University, both are top ranking private university in Bangladesh. Dr. Khan has a large number of publications in national and international books, journals and periodicals. His research interests are: disaster, environment, climate change, globalization and women empowerment.

Acronyms and Abbreviations

ADB	Asian Development Bank
BANBEIS	Bangladesh Bureau of Education, Information and Statistics
BBS	Bangladesh Bureau of Statistics
BCCSAP	Bangladesh Climate Change Strategy and Action Plan
BMD	Bangladesh Meteorological Department
BMP	Bangladesh Mohila Parishad
BRAC	Bangladesh Rural Advancement Committee
CATAW	Coalition for Assessing Tsunami Affected Women
CCDMC	City Corporation Disaster Management Committee
ccGAP	Bangladesh Climate Change and Gender Action Plan
CDA	Community Development Association
CDM	Comprehensive Disaster Management
CDMP	Comprehensive Disaster Management Programme
CDR	Crude Death Rate
CEDAW	Committee on the Elimination of Discrimination against Women
CPD	Centre for Policy Dialogue
CPP	Cyclone Preparedness Programme
CRED	Centre for Research on the Epidemiology of Disasters
DCM	Disaster Crunch Model
DDMC	District Disaster Management Committee
DFID	Department for International Development
DM	Disaster Management
DMB	Disaster Management Bureau
DMRD	Disaster Management and Rehabilitation Division
DRM	Disaster Risk Management
DRR	Disaster Risk Reduction
FAO	Food and Agriculture Organisation
FFWC	Flood Forecasting and Warning Centre
FGD	Focus Group Discussion
FYP	Five Year Plan
GHG	Greenhouse Gases

GO	Government
GoB	Government of the People's Republic of Bangladesh
GOs	Government Organisations
HEC	Human Ethics Committee
HFA	Hyogo Framework for Action
HSBC	Hongkong and Shanghai Banking Corporation
IFRC	International Federation of Red Cross and Red Crescent Societies
IMDMCC	Inter-Ministerial Disaster Management Coordination Committee
IPCC	Intergovernmental Panel on Climate Change
ISDR	International Strategy for Disaster Reduction
LDC	Least Developed Country
MDG	Millennium Development Goal
MMR	Maternal Mortality Ratio
MoDMR	Ministry of Disaster Management and Relief
MoFDM	Ministry of Food and Disaster Management
MoWCA	Ministry of Women and Children Affairs
MP	Mohila Parishad (Council for Women)
NAPA	National Adaptation Programme of Action
NDMC	National Disaster Management Council
NGO	Non-Government Organisation
NP	Nari Pokko (An organisation for women)
NPDM	National Plan for Disaster Management
PAR	Pressure and Release
PDMC	Pourashaba Disaster Management Committee
PIC	Project Implementation Committee
PIO	Project Implementation Officer
SDG	Sustainable Development Goals
SMC	School Managing Committee
SOD	Standing Orders on Disaster
SPCR	Strategic Programme for Climate Resilience
SPSS	Statistical Package for Social Sciences
TR	Test Relief
UDMC	Union Disaster Management Committee
UNCBD	United Nations Conventions on Biodiversity
UNDAW	United Nations Division for the Advancement of Women
UNDP	United Nations Development Programme
UNFC	United Nations Framework Convention
UNFCCC	United Nations Framework Convention on Climate Change
UNICEF	United Nations International Children's Emergency Fund
UNO	Upazila Nirbhahi (executive) Officer
UNWFP	United Nations World Food Programme
UP	Union Parishad
USAID	United States Agency for International Development
UzDMC	Upazila Disaster Management Committee
VGD	Vulnerable Group Development

VGF	Vulnerable Group Feeding (long-term aid for the extreme poor)
WEDO	Women's Economic Development Outreach
WFW	Women for Women
WHO	World Health Organisation

List of Figures

List of Tables

Chapter 1
Introduction: Gender and Women in Understanding Disasters

1.1 Introduction

Bangladeshi women have been the most vulnerable group in the country during different natural hazards such as the 1988 flood and cyclones in 1970, 1991, 2007, 2009 and many others. Regardless of caste, class, ethnicity or gender, disasters appear to be gender-neutral events affecting people and their livelihoods; however, researchers report that impacts are considerable on poor communities and mostly on women (Akhter, 1992; Kabeer, 1990; Kabir, 1992; Khan & Azad 2012; Khatoon, 1995; Khondker, 1996; Nasreen, 1995, 2008, 2010, 2012; Saha, 2015).

During a disaster, gender-based disparities have a higher degree of impact than normal because of the breakdown of social and family order, institutional security and protection. Baten & Khan (2010) reported that women are the most marginalised and vulnerable people to disasters because of a discriminatory socio-economic status and unequal access to Bangladeshi resources. From a gender perspective, women in the developing world have been and are still disproportionately affected by disasters (Ahmad, 2015; Alam & Rahman, 2014; Denton, 2000, 2002; MacGregor, 2010), whereas women in rural areas are extremely vulnerable to disasters in a developing country (Dankelman, 2002; Nelson et al., 2002; Terry, 2009). The disproportionate and unequal vulnerability of women is linked to unequal access, lack of entitlements/control of resources and properties, discriminatory gender relations and gender differences in response to disasters (Kumar-Range, 2001; Terry, 2009). The lack of rights to land and resources, as well as limited access to credit (Agarwal, 1992), in addition to socio-cultural norms and practices, discriminatory gender roles, limited mobility and less contribution to the decision-making power at household and community level makes women more vulnerable (Terry, 2009).

The extremely common perception about women is that they are dependent, weak, incapable and subordinated; moreover, they are considered as a burden that keeps them isolated from the mainstream processes of society such as planning and decision-making. Women comparatively have lesser access to warning information about cyclones; in multiple cases, they are dependent on the decision-making of male

M. M. Khan, *Disaster and Gender in Coastal Bangladesh*,
https://doi.org/10.1007/978-981-19-3284-7_1

family members. They wait for their husbands to return home and expect husbands will take them to a shelter centre provided by GO on higher land or other safe places. Sometimes, they are unwilling to leave the house because of their desire to protect assets and a fear of the poor management of shelter centres. The shelter centres are not well designed and insensitive to gender and culture-specific requirements. Many men and women live in the shelter centre; moreover, shelters lack separate toilets, water, shower systems and toiletries such as sanitary pads for women. They are unpleasant particularly for pregnant and young women. Moreover, in a culture placing a high premium on women's modesty, the dress code *saree* can become a death trap as it is difficult for women to quickly move in them. At the time of relief distribution after a disaster, women sometimes struggle to collect relief and other aid. Finally, when women manage to get relief and other aid, the male heads of households may sell relief items to meet their other demands (such as cigarettes) and priorities rather than household requirements (Nasreen, 2012).

However, during the post-disaster period, when rehabilitation programmes such as land distribution and house-building materials start, women and women-headed households once again can become marginalised from acquiring such assets. For many cases, land, house-building materials and houses are allocated/registered to the male victims of the disaster rather than the wife because of the law, policy and socio-cultural values of the society.

Women are competent to help family cope better during a disaster; moreover, they are agents of change too (Dasgupta, 2010; Nasreen, 2008). However, their skills, abilities and life experiences may not be identified as resources. Despite the GO rhetoric that increased vulnerability and dependency of women, they are not remarkably involved in risk reduction, disaster preparedness and relief or recovery efforts, although the situation improved after incorporating women in disaster management. Women could be leaders and contribute to a disastrous situation because they have the experience and knowledge of nature and environment because of close attachment. Women are proactive during emergencies. During disasters, they take the responsibility for all family-related tasks and guide family members for alleviating distress.

For post-disaster migration, gender disparity is observed, e.g., in many cases, decisions about migration were taken by male members of households and extremely few women decided to independently migrate. Studies demonstrate that disasters have resulted in massive community displacement either temporarily or permanently because of the lack of livelihoods, jobs and income opportunities from their own community (Mallick et al., 2011; Mallick & Vogt, 2012, 2014). Both women and children experienced inhuman situations and were the worst victims of vulnerability during the post-disaster period when the migration of males for livelihoods left them behind (Mallick et al., 2011). Migration resulted in exclusion from their own community, inclusion in a new society, and threw displaced women in high vulnerability and socio-culturally constrained situations.

Existing data and study in Bangladesh has confirmed that disaster, women and vulnerability discourse leads to the stereotype that "women are helpless victims" for all time periods (Nonoguchi, 2012) and simplifies that women are marginalised and vulnerable during a disaster. Women are considered to always be dependent, i.e., a burden on society and the source of problems. They are perceived as being capable of work only in the household or fulfilling care-giving responsibilities; therefore, women are usually viewed as more vulnerable because they are poorer (Cannon, 2002; Denton, 2000, 2002; Nelson et al., 2002). The positive aspects of women's coping capacity, adaptive capacity, leadership in disaster management, knowledge, skill, experience and contribution are often ignored by family and community; however, women and vulnerability are not synonymous (Masika, 2002; Terry, 2009). The poorest women have their own coping techniques because of their everyday survival experience. In addition to their survival experiences, women are mainstreamed, empowered and more focused on by GO and NGOs; consequently, the bargaining power/agency of women has increased, e.g., the participation and inclusion of women in the decision-making process at national, regional and local levels is ensured by the different policy (climate change and disaster management policies) of GO and NGOs.

This study focuses on the requirements of vulnerable coastal women and provides a voice of the critically affected women heard, as well as describe their experiences and crises in the coastal area in Bangladesh during a disaster and make them an important part of the global discussion. How do the existing social structure and multiple institutions function in coastal Bangladesh during a disaster? This study analyses whether women are vulnerable only because they are women or is there something else such as a socio-economic status? Moreover, in what manner do women cope and adapt? This study focuses on women involvement with decision-making processes at the local level, as well as the development of their agency, their role as leaders in disaster management and how this involvement impacts disaster management processes at a local level. Are there any changes in the socio-cultural practices as well as in the role and responsibilities of women in the coastal area because of external interventions such as GO and NGOs? This study identifies multiple approaches by which women are vulnerable and resilient and their capacity to take more leadership to manage disaster and to contribute to the development process. Using a mixed method approach, this study addresses the potentiality of women's leadership in disaster management if there is a favourable atmosphere in the locality. Ensuring the involvement and participation of women with development activities and with different policies led to changes in daily practices, additional changes are possible if women are mainstreamed and empowered by implementing existing disaster-related policies.

This study explores local practices, the nature of access and entitlements to resources, the vulnerability of female-headed households, women's involvement and leadership in disaster management, their adaptive mechanisms, changes in everyday practices in the locality and the role of GO and NGO in the development of the agency/bargaining power of women. This study provides guidance and a constructive commentary for policy-makers to take gender-neutral initiatives in the development and disaster management process in future and programmatic intervention in the Bangladeshi coastal region.

1.2 Rationale of the Study

Disaster-related vulnerabilities are not equally distributed all over the world because of variations in economy, culture, development processes, settlement processes, politics and resource management. This is certainly placing many people at higher risk than others. Cannon (as cited in Enarson & Morrow, 1998) argues that the experience of environmental hazards and risks of catastrophic disaster are shaped by the elements of social structures, including class, religion, age, physical ability, and gender. Different social groups and different individuals carry disproportionate "vulnerability bundles", e.g., women are the most at risk when hazardous conditions unfold in disastrous events such as cyclones (Ikeda, 1995). Gender, accordingly, is a pervasive division that provides access to social and economic resources away from women. Disaster is an event that opens out natural practices, traditions, gender, relationship and power relations in a community even in intimate relationships. For example, Akhter (1992) reports that, in the struggle to survive during the disaster, for a Bangladeshi father of five daughters and a son caught up in a cyclone and flood, the father released his daughters one after the other such that his son could survive.

The social experiences of disaster confirm, reflect, disrupt and otherwise engage gendered social relationships, practices and institutions (Enarson & Morrow, 1998). The care-giving responsibility of women increases women's vulnerability during a disaster that Bari (1992) describes as coping with poverty being tougher for women because it hits women the hardest. Men may lose their fishing equipment, which is important to earn a living, or their children die or their homes and belongings are washed away. However, at the end of the day, it is their wife or mother who has to cook to enable the remaining people to survive. Therefore, the vulnerability of women is a historical and cultural phenomenon and economic vulnerability makes women dependent on assistance such as relief where access to that assistance is differentiated because of gender, marital status, age and personal communication, e.g., the vulnerability of the single mother, widow or frail elderly woman is particularly evident, reinforcing the familiar notion of women as prototypic disaster victims and dependents, even inducing emotional dependence (Enarson & Morrow, 1998), for disaster managers.

In the context of Bangladesh, this study is extremely significant for many reasons that will contribute to the theoretical underpinnings and increase the horizon of knowledge in the disaster, women and vulnerability discourse. In recent years, in the disaster-prone coastal area of Bangladesh, both GO and NGOs have taken certain development initiatives. They are focusing on and prioritising women in disaster management and have started to recognise that the involvement of women in decision-making processes and earning projects make projects and disaster-related initiatives more effective. Both GO and NGOs have considered initiatives and different training programme strategies to cultivate additional gender-equitable attitudes (Nazneen et al., 2011; Nazneen & Sultan, 2012; Odendaal, 2001; Panday et al., 2016; Peacock, 2003; Uddin et al., 2015; Varga, 2001). However, the lack of recent empirical data and results related to women's role and involvement is limiting their initiatives to certain extent. This study will definitely fill a gap and lack of theoretical basis on the role of women in disaster management if they are incorporated in a disaster management position. Thus, in the coastal area of Bangladesh, it will improve the efficacy to take sustainable initiatives to encourage gender-neutral attitudes.

In Bangladesh, patriarchal norms are dominantly influenced and guided via traditional gender ideas and expectations (Miescher, 2005) and economic marginalisation (Connell, 2006). This study provides an improved understanding of the vulnerability of women focusing on unequal distribution, lack of entitlements and discriminatory access to resources of women as institutional practices of society, in addition to coping mechanisms of women, particularly female-headed households and women's leadership. This study contributes to the understanding of gender-based identity by addressing changing trends in practices in the coastal area in terms of the involvement of women with the development process, earning/income-generation activities, decision-making process, entitlements to resources/wealth/income and employment opportunities. This study shows the socially constructed nature of masculinities (Adu-Poku, 2001) that required to be removed by implementing the women rights focused on different policies and action plans.

This study is important because it considers that women are agents of change; they have started to go out from their traditional role and responsibilities. Women in coastal areas are involved with outdoor income-generating activities; their contribution and involvement with outdoor activities is appreciated and allowed by male members of family and society. This study focuses on the nexus between changes in the practice of community and entitlements and access to resources and income-generating activities. By integrating both quantitative and qualitative methodology, this study attempts to avoid pitfalls by involving women at a household level via face-to-face interviews for obtaining quantitative survey data. Moreover, to obtain insight and qualitative data, focus group discussions and in-depth individual interviews have been conducted because women and men contribute from their own position in a society. After this study, there is an opportunity to build and develop the women's capacity to empower women and develop disaster management leadership for an uncertain future in the coastal area. In-depth individual interviews were undertaken with different stakeholders to understand how programmes in the coastal area could be implemented for mainstreaming women and their emancipation/empowerment.

1.3 Hazards and Disasters: Global Context

Over the last few decades, the world has experienced extremely frequent/unusual natural and technological disasters that have massively impacted the lives and livelihoods of people all over the world. IFRC (2010) records that, in 2013, 330 natural disasters resulted in an overall death toll of 22,452 and affected > 96.5 million people with an estimated economic damage of $118.6 billion worldwide. Studies (Dasgupta et al., 2010, 2012) report that the number of disasters increased on an average 6%

every year between 1962 and 1992 and annually affected 200 million people during the 1990s; moreover, from the late 1960s, there was a fourfold increase in disasters.

Although there are disasters occurring with varying frequency and level of impact all over the world, Asian nations held the top position as the most disaster-afflicted region in the world between 1975 and 2006 (Saha, 2015). ADB (2010) reported that, during this time (1975–2006), 89% of people in Asia were affected and 44% of total damage were economic. In 2008, because of the cyclone *Nargis* in Myanmar, > 80,000 people died and ~ seven million people were affected. The economic loss was > US$10 billion and ~ 1 million people were compelled to live without sufficient safe shelter, food, medical care and drinking water (UN, 2009 as cited Saha, 2015).

The Indian Ocean earthquake and tsunami in 2004 caused > 220,000 deaths (Kron, 2012) and displaced 500,000 persons in Sri Lanka itself. In Sri Lanka after the tsunami, incidents of sexual harassment, such as rape and gang rape; physical abuse of women and girls; and a sense of insecurity and fear were reported in relief camps (Chadwick, 2005). Fisher (2010) argues that the lack of security and chaotic conditions in shelter centres created an atmosphere in which girls and women were vulnerable to sexual violence.

In Florida, after Hurricane Andrew, the rate of violence increased in 1991 and there was a 50% increase in spousal abuse calls, including family violence, spousal abuse, child abuse and domestic violence (Morrow, 1997; Shaw et al., 1995; Wilson et al., 1998). In 2005, Hurricane Katrina caused 1,836 fatalities, made 1,000,000 homeless, resulted in 7,500 cases of health-related illness and injuries and caused economic damage of > US$ 125 billion in New Orleans and Mississippi. Furthermore, it exposed 80% of people to extreme flood and inundation (Brakenridge et al., 2012; Kron, 2012; Tierney, 2006).

Increase in disasters, including cyclones, floods, earthquakes and rising sea levels all over the world, is causing the drying out of wetlands, irrigation crises, threats to coastal people, and imbalance in coastal ecosystems, destruction of fisheries and infrastructure, increased death tolls and other economic losses. Flood disasters itself resulted in a death toll of 175,000 worldwide and affected 2.2 billion people between 1975 and 2002 (Jonkman, 2005 as cited in Saha, 2015) and cost ~ US$ 470 billion (HSBC, 2011). As reported by the secretariat of ISDR (2004), in the last ten years, 478,100 people were killed, and > 2.5 billion people were affected and ~ US$690 billion of economic losses have been attributed to flooding. Mental depression, distress, disparity and discrimination, unemployment, disability, trauma, illness, dislocation of social networks, infectious diseases, shortage of drinking water, sanitation, malnutrition and disruption of health care are increasing in the society because of disasters. Events were trigged by hydro-meteorological hazards in such a severe manner that 97% of total people were affected and 60% have had economic losses (Shaw, 2006).

The magnitudes of effects from a disaster always vary depending on class, gender, age, race and ethnicity (Saha, 2015). Even in wealthy industrialised countries, people in the poorest regions of the world as well as the most vulnerable groups experience double challenges because of social exclusion, type of residence and scarce access to resources (Fothergil & Peek, 2004; Peacock et al., 1997). Disaster experts report that women, children, disabled and elderly people are extremely vulnerable groups to face the adverse effects of catastrophic events (Cutter et al., 2003; Peek & Stough, 2010 as cited in Saha, 2015). Both response and recovery of women varies from that of men because of different experiences during disasters (Fothergill, 2004). In particular, female-headed households and single mothers face vulnerabilities because of the structural positions in society as well as their responsibilities and routine activities (Tobin-Gurley et. al., 2010); pregnant women are less likely to have access to adequate family and medical care (Zahran et al., 2013 as cited in Saha, 2015). Furthermore, they disrupt social arrangements and inflict challenges to recovery, environmental remediation and mitigation of future disasters. In addition to these limitations, disasters fuel up social disorder, crime and domestic violence (Saha, 2015; Tierney et al., 2006; Zahran et al., 2009); break up family relations (Peek et al., 2011 as cited in Saha, 2015); and cause uncertainty, maladjustment to new places and environment, sadness, loneliness, marginalisation of different groups and income and wealth inequality (Nel & Righarts, 2008 as cited in Saha, 2015).

1.4 Hazards and Disasters in the Coastal Area of Bangladesh

Bangladesh has been identified as the country at the highest risk for disasters. It leads the top ten countries in the Asia–Pacific region for floods, fifth for storms and eighth for earth quakes because of its physical proximity (Asia–Pacific Disaster Report, 2010). For natural hazards, Bangladesh ranks as one of the most vulnerable countries, particularly to increases in sea level, because of its geographical location, land characteristics, multiplicity of rivers, monsoon climate and climatic variability (Huq & Karim, 1999; Zaman, 1999). The World Bank (2006) reports that Bangladesh will be a victim of environmental catastrophe because of higher sea levels. Note that ~ 33 million people will be evicted because of permanent flooding in multiple areas of coastal Bangladesh. By 2030, an extra 14% of Bangladesh will become vulnerable to floods, and the current vulnerable areas will increasingly experience severe flooding. The vulnerability of Bangladesh to climate change is considered as massive and disastrous, which places Bangladesh as one of the most susceptible countries in the world (ADB, 2004).

The scale of climatic disasters that Bangladesh has been facing for a long time period will continue because of the rise in sea-level geography of the coastal area, the high density of population and a lack of coastal protection, in addition to its location at the head of the triangularly shaped Bay of Bengal. Note that ~ 40% of total global storm surges and deadliest cyclones, in terms of deaths and casualties, in the past 50 years have been recorded in Bangladesh. Moreover, ~ 10 million Bangladeshi citizens are annually impacted by natural hazards (Khan & Kamal, 2013). CRED (2015) estimates that between 1979 and 2010, over 234 disasters occurred; they caused the death of 191,836 people and directly affected 229 million people badly and caused economic damage of ~ US$5.6 billion. Studies report (Huq et al., 2004; Nahar et al., 2014) that Bangladesh, as a developing country, has multiple challenges, including rapid population growth, illiteracy, and poverty. Its economic development, along with disasters, particularly cyclone, floods, coastal bank erosion, rising sea levels and tidal surges, challenge its economic development. Note that 25% of Bangladesh is flooded; moreover, every 4–5 years, ~ 60% of the total area is flooded (GoB, 2010). The deadliest disasters in Bangladesh over the last three decades are listed below (Table 1.1).

Out of the 64 districts of Bangladesh, the coastal zone covers 16 districts (an area of 47,211 km^2) or 32% of the country's geographical area (where 35 million people), i.e., 28% of the country's total population live in 6.85 million households (BBS 2011; Huq et al. 2004), which are extremely vulnerable to climatic hazards. In the Bangladeshi coastal area, there have been 70 major cyclones in the last 200 years. Moreover, in the last 35 years, ~ 900,000 people have died because of these disasters (Mallick & Vogt, 2009). For socio-cultural and political reasons, the Bangladeshi coastal area differed from the rest of the country, which often limits the access of people to endowed resources. Both socio-cultural and political features of the coastal area perpetuate risks (Shamsuddoha & Chowdhury, 2007). The higher number and intensification of disasters every year, including cyclones, floods, storm surges and tidal surges, in addition to higher sea levels, makes the livelihoods of coastal people vulnerable. Furthermore, changes in upstream river discharge, lack of strong embankments, water logging, flash floods upstream, shortages of proper canals and sluice gates for controlling the water, the presence of salinity and unplanned shrimp cultivation are making the life of coastal community harder and more exposed to vulnerability. During pre-monsoon and post-monsoon seasons, violent storm surges and tropical

Table 1.1 Major cyclones that hit the Bangladeshi coast

Year (Storm name, where given)	Maximum wind speed (km/h)	Maximum storm surge height (metres)	Death toll
11 May 1965	161	7.6	19,300
15 December 1965	217	3.6	870
01 October 1966	139	6.7	850
12 November 1970 (Bhola)	224	10.0	300,000
25 May 1985	154	4.6	11,000
29 April 1991 (Gorky)	225	7.6	139,000
19 May 1997	232	4.6	150
15 November 2007 (*Sidr*)	223	4.0	3,360
15 May 2009 (*Aila*)	92	3.50	190
9 November 2019 (Bulbul)	120	3.0	25
19 May 2020 (Amphan)	160	3.0	26

Source National Plan for Disaster Management 2010–2020

cyclones are extremely common; therefore, the coastal zone of Bangladesh is known as a geographical ‘death trap’ (Shamsuddoha & Chowdhury, 2007).

In the last decade, cyclone *Sidr* hit the Bangladeshi coastal area on 15 November 2007. As cyclone *Sidr* approached, 40,000 Red Crescent volunteers were deployed to direct residents from these 15 affected districts. There were ~ 3,406 confirmed deaths and > 55,000 people sustained permanent physical injuries (Nahar etal., 2014; Paul, 2009), in addition to economic loss, and damage to houses, crops and vegetable lands, wood and fruit trees, domestic animals, community infrastructures and coastal defences (Haq et al., 2012; Saha, 2015). Note that 10 million people were affected, 1.5 million houses were damaged (Paul, 2009) and 186,883 hectares of crop areas were completely damaged (GoB, 2010).

On 25 May 2009, cyclone *Aila* severely hit the south-western coastline of Bangladesh and neighbouring districts of India; the poorest of the poor were the worst victims (Onneshan, 2009). Compared to cyclone *Sidr* that struck the Bangladeshi coastal area in 2007, *Aila* did not result in a high death toll; however, the cyclone’s aftermath was horrendous as half a million people migrated because of permanent water-logged conditions in the affected areas (Onneshan, 2009); furthermore, the storm displaced ~ 2 million people, most of whom returned to their places of origin and started spontaneous and assisted recovery. The catastrophic cyclone *Aila* killed 190 people along with ~ 8000 probably dead (missing/never found) persons (Mallick & Vogt, 2012), thus affecting > 3.9 million people, including livelihood disruption and infrastructure destruction (IFRC, 2010). Note that ~ 352 unions and 62 *upazilas* in 14 districts were affected (Mallick & Vogt, 2012). Cyclone *Aila* damaged 68,385 acres of crop land, and partially damaged 239,212 acres and killed > 1 million livestock (including cows, goats and chickens, all of which were people’s subsistence and their means to cash), destroyed/washed away shrimp farms and homestead gardens depriving people both of their prime income source and non-economic revenues and assets (Mallick & Vogt, 2012). Moreover, in these 14 districts, ~ 258.5 km of roads were extremely damaged to be used and 509 km of embankments were destroyed and 2,155 km were damaged (Kumar et al., 2010 as cited in Mallick & Vogt, 2012).

Under the influence of the “very severe” cyclonic storm Amphan (19 May 2020), Feni, Chattagram, Satkhira, Khulna, Bagerhat, Jhalokathi, Pirojpur, Borguna, Patuakhali, Bhola, Barisal, Laxmipur, Chandpur and their offshore islands and chars were inundated by a storm surge of 4–5 feet height above the normal astronomical tide. The coastal districts experienced wind speeds of up to 140–160kmph in gusts/squalls with heavy to extremely heavy falls during the storm’s passage. After the great danger signal and evacuation order of the Government of Bangladesh (GoB), > 2.4 million people were moved to 14,636 permanent and temporary shelters in 19 coastal districts. During this time period, the number of shelters increased considering the physical distancing norm and COVID-19 guidelines.

Cyclone Amphan affected > 1 million people in nine districts in Khulna and Barishal divisions of Bangladesh on May 20. UNDP (2020) reports that 26 people were killed, 55,667 houses were completely damaged, and ~ 162,000 partially damaged. Note that ~ 149,000 hectares of agricultural land and fish farms worth about BDT 3.25 billion were damaged, and 150 km of protection embankments were washed away at 84 points in 13 districts; moreover, 200 bridges and culverts and 100 km of roads were damaged. Note that ~ 15 million clients lost electricity, and multiple fresh water ponds within forests were flooded with seawater. The preliminary estimates of total damage cost caused by the cyclone is BDT 11 billion (Map 1.1).

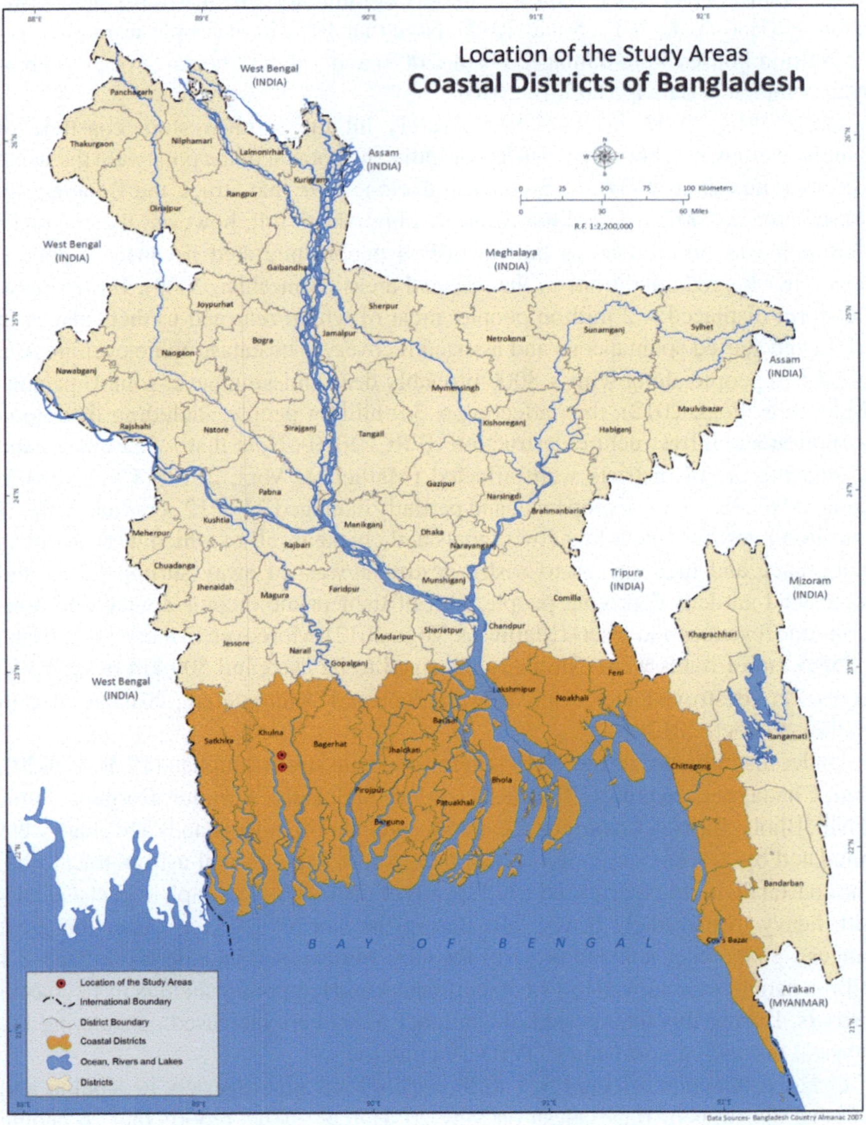

Map 1.1 Coastal districts of Bangladesh

1.5 Disaster, Gender and Women in Bangladesh

"Disaster", "gender", "women", "vulnerability" and "change in women's lives at a community level are an overriding concern in the Bangladeshi coastal area for developing the discourse, as well as for developing a gender-neutral disaster management system. This section provides a brief literature-based introduction, including the socio-economic and demographic context of women in Bangladesh, the nature of vulnerability of women and coping and adaptation mechanisms from gendered perspectives. Secondly, in the context of gender relations in Bangladesh, it explores disaster policies and institutional arrangement in Bangladesh. Thirdly, for empowering women, the section discusses literature related to the role of GOs and NGOs, position of women in disaster management and climate change policies, position of women disaster management institutions and different programmes for improving women's leadership in disaster management. Finally, to develop a theoretical and conceptual framework, it focuses on relevance, updating and consistencies of theories.

From a broad perspective, each section of the literature has been considered as a background to explore how disaster, gender, women and vulnerability function during a disaster in the Bangladeshi coastal area. For example, literature related to vulnerability, coping and adaptation mechanisms of women is discussed to analyse local practices, as well as unequal access and entitlement to resources that intensify disproportionate vulnerability for women. In this chapter, initiatives by GOs and NGOs are discussed to understand the participation of women/leadership during disaster risk and vulnerability management (DRM). The development of a women-based agency has been addressed to explore how interventions of GO and NGOs contribute in mainstreaming women with decision-making processes and attempts to explore whether there is any change in local practices, lifestyle and livelihood of women at a community level.

1.5.1 Disaster and Women: Socio-Cultural Context in Bangladesh

In Bangladesh, the population size has been continuously increasing, although the annual growth rate of the population is decreasing. The population size doubled between 1974 and 2011. Note that most of the population in Bangladesh is in the 15–49 age group, and children under 15 years and adults over 60 years, who constitute the dependent population, have gradually decreased over the decades since 1981 (BBS, 2018). However, the population growth would not be a problem if everyone had access to education, quality food, healthcare facilities, vaccination facilities, pure drinking water, safe housing, proper advice and opportunities for family planning.

GO policies promote the empowerment of women and gender mainstreaming, integrating gender issues in the planning and budgetary process, strengthening women's participation in decision-making and increasing employment opportunities for women. The sex ratio (m/f 100) shows that the difference between women and men has moved towards equality; it has declined from 104.7 in 2012 to 100.2 in 2017 (BBS, 2018). The decrease is evident in both rural and urban areas; however, in urban areas, the sex ratio is considerably higher; it declined from 103.0 in 2013 to 100.1 in 2017 because of a high rate of male migration compared to female migration. At national level, 86.7% of households were headed by men in 2011 (BBS, 2018); there are variations in the ratio of households headed by men and women; however, male dominance exists in both urban (13.8%) and rural (14.4%) areas over time.

Bangladesh signed the declaration "Education for All" at the world conference in 1990 in Jomtein, Thailand (ADB, 2004). Despite multiple strategies to educate children and women, Thailand still faces multiple challenges to reach a satisfactory benchmark of education and literacy. The literacy rate of women (>7 years) has significantly increased from 53.3% in 2012 to 70.2% in 2017 over time; however, the literacy rate of men increased from 59.2% in 2012 to 74.3% in 2017 (BBS, 2018). Focused efforts and collaboration with different development partners and organisations have led to considerable improvements in the literacy rate of adults aged 15 + years. In 2017, the adult literacy rate of women in urban areas was 78.4% compared to an adult literacy rate of 63.2% in rural areas (BBS, 2018).

To provide energy and enthusiasm for life, optimal and balanced nutrition for good health is unquestionable and important for the fight against diseases and ensuring correct balance in the body. In Bangladesh, although women are generally considered as the inferior sex, the GO has considered women's health as one of the indices of human development in most rural families. Despite considerable improvement in the health sector over the last few years, malnutrition among women and children, including maternal mortality rate, is still extremely high in Bangladesh (World Bank, 2008). CDR by sex at a national level was 5.8 for men compared to 4.4 for women per 1000 population in 2017; similar trends were observed in both rural and urban areas (BBS, 2018). WHO (2001) estimates that, because of women's reproductive responsibility, domestic chores responsibility and stress for earning, the morbidity rate for women is higher in Bangladesh than for men, which possibly makes them more susceptible to health hazards compared to their male counterparts (Begum & Sen, 2009).

Child mortality rates for both girls and boys have considerably declined after 1981 in Bangladesh. While child mortality was high for girls compared to boys till the early 1990s, it may have reversed after the mid-1990s (BBS, 2018). Under five child mortality rates halved over the last 25 years in Bangladesh from 144 per 1000 in 1990 to 38 per 1000 in 2015; however, the target was 48 per 1000 in 2015 (UNICEF, 2015). Over the last few decades, child survival rates made remarkable progress because of the successful implementation of an immunisation programme, control of diarrhoeal diseases, vitamin A supplementation, dietary awareness and balanced food for mothers and children.

In Bangladesh, MMR has nationally decreased from 6.48 in 1986 to 2.16 in 2010, although it was a relatively higher decline in urban areas (BBS, 2018). Although Bangladesh has progressed and MMR has remarkably declined, the rate of decline

Table 1.2 Labour force aged 15 years and over by sex and residence area (millions)

Year	National		Urban		Rural	
	Women	Men	Women	Men	Women	Men
2005–2006	12.1	37.3	2.8	8.9	9.3	28.4
2010	17.2	39.5	4.0	9.3	13.2	30.2
2013	18.2	42.5	5.1	12.0	13.1	30.5
2015–2016	19.1	43.1	4.8	12.5	14.3	30.6
2016–2017	20.0	43.5	5.0	12.9	15.0	30.7

Source Gender Statistics of Bangladesh, 2018, BBS

is not at the desired level (BBS, 2008) because of post-partum haemorrhaging, eclampsia, obstructed labour and unsafe abortion (BBS, 2010).

BBS (2012) defines that labour force comprises people aged 15 years and over who are economically dynamic and who can work for producing goods and services. At a national level, the male labour force increased from 42.5 million in 2013 to 43.5 million in 2016–17 while the female labour force increased from 18.2 million to 20.0 million during the same period. In both urban and rural areas, the rate of women's participation increased at a faster rate than that of men (Table 1.2).

Between 2005–2006 and 2010, the annual growth rate of labour force for women was 8.69% compared to 1.40% for men at a national level. The growth rate of labour force for women in rural areas was considerable compared to the growth rate of labour force for women in urban areas (8.52%) compared to 1.50% of men (BBS, 2012). In rural areas, the female labour force has quickly increased compared to the male labour force because of the free education for girls, priority and quota of women for new employment opportunities, priority for women in different GO and NGO projects and programmes, micro-credit loans for women, loans for small entrepreneurs in rural areas, growth of the garment industry and the national economic growth of Bangladesh.

Bangladesh has made exemplary developments in poverty reduction over the last decade and implemented livelihood programmes for rural poor and ultra-poor, created employment opportunities, provided different training, provided unconditional loans or financial support and was promoted to a lower middle-income country in 2015 with the per capita income rising from $407 in 2007 to $1041 in 2015. The upgradation of Bangladesh to a lower middle-income country has been particularly assisted by large foreign remittance for males and females, garment trade and increase in employment (http://data.worldbank.org/country/bangladesh). Bangladesh has, however, experienced greater progress in increasing education, managing population growth rate, increasing employment and earning opportunities, health and sanitation, child and maternal mortality and other sources of human development compared to reduction in income poverty itself (UNWFP, 2006).

However, the poverty of women in Bangladesh is caused by the monthly average income, health status, nature of paid and unpaid jobs and employment opportunities. In terms of access and ownership of resources, women are vulnerable to poverty because they are directly and indirectly discriminated against households (Nawaz, 2009). The average monthly household income of male-headed households in 2010 was BDT.11763 ($USD143), which is 21% higher than that of female-headed households at BDT. 9725 ($USD 119). Similarly, the average monthly expenditure for male-headed households in 2010 was BDT.11346 ($USD 138), which is 27.8% higher than that for female-headed households at BDT 8874 (BBS, 2018), thus demonstrating that women save more, although they have a smaller income.

Women empowerment includes the involvement of women with social, political and educational opportunities, which has been increasing since 1971. The inadequacy of political participation by women was an important barrier to women's empowerment; furthermore, different measures for its removal have been taken by GO, NGOs, women's organisations, civil society and international NGOs over the period. Bangladesh made formal commitments for promoting the political participation of women by adopting important international strategies initiated by the UN, thus reserving quota at both local and national elections, promoting direct participation of women, ensuring proportional participation of women, and increasing economic participation and emancipation of women.

The recent amendments to Union Parishad Ordinance (1997) on the provision of 33% of the reserved seats in UPs for the direct election of the women changed the situation of women's representation at a local level (CPD, 2000; Rahman, 2014). For example, the number of women elected in the general assembly of the country in 1973 was 4.8% of the total number of elected members; it has increased to 20.9% in 2017 (BBS, 2018). In the UP, the participation rate of women as chairman in 2012 was 0.0% and 0.7% in 2017, which is extremely small compared to male participation (100% in 2012 and 99.3% in 2017) and compared to the total number of women in the country (BBS, 2018).

In addition to the direct participation of women in politics, gender quotas (30% at local election and 17% at national election) for women's representation in legislative bodies have been confirmed by the Local Government Act of Bangladesh. The total number of chairmen at the upazila level is extremely poor; it was only 0.6% in 2012 and slightly increased to 1.4% in 2017 (BBS, 2018). The participation of women in civil services in 2012 was 17.97% and in 2017 was 30%, which shows that participation by women has proportionately increased.

In terms of the decision-making role, 47% of decisions with respect to family expenditure are undertaken by the husband; only in 32.6% of cases, the decision is jointly undertaken by the mother and other household members. However, for cooking food, 50.9% of decision-making is by the mother and only 27.1% by other household members. For attending to health requirements, only in 5.6% of cases can the mother make a decision to visit a health centre for her own health by herself compared to 39.1% decisions by the husband and 40.7% of joint decisions (BBS, 2018).

In this study, the under-representation of women at all levels is addressed as one of the major reasons for excluding women from decision-making processes (Aminuzzaman, 2010 as cited in Alston, 2015). For example, the overall direct representation (excluding quotas) of women in national-level politics is only 2%–3% (Alston, 2015), although the Prime Minister, Leader of the Opposition, Speaker of the Parliament and Former Prime Minister are women.

Despite the recognition of equal rights for all, irrespective of sex, religion and caste by the Bangladeshi constitution, women have been most vulnerable to violence in Bangladesh. Violence against women includes rape, acid throwing, physical torture and physical and mental torture for dowry. The trend of rape has fluctuated with time; it sometimes increases and sometimes decreases. After 2003, the incidents of rape has reduced in 2004, 2005 and 2006 but increased in 2007. In 2009, the records of rape case demonstrated a reduction at 2973 incidents; however, in 2010 and 2011, it started to increase again with the number of incidents being 3367 and 3675, respectively. The situation of rape violence has been deteriorating rather than improving.

Women's oppression for dowry continues with 4061 incidents reported in 2009, 5331 in 2010, and 7079 incidents in 2011. Gradually, violence for dowry increased over the 2009–2011 period (BBS, 2012, 2018). However, after 2012, other violence against women reported in Bangladesh has improved, which is reflected in Table 1.3, although not all types of violence against women are consistently reducing but are rather fluctuating. Some violence against women, i.e., women trafficking, acid throwing and stalking, are consistently decreasing. This is attributed to improve monitoring by law enforcement agencies; use of the latest technology to identify criminals/criminal activities; regular monitoring of cyber and social media; updated judicial systems; and laws, speedy and special trials for certain specific crimes such as rape, and specialist GO units to deal with specific crimes such as trafficking. Moreover, there are education programmes such as inclusion of issues such as trafficking, dowry and acid throwing in school textbooks; massive sensitising and awareness programmes for gender equality at a grassroots level; campaigning in electronic and print media for women's rights; the active role of media and corporate organisations; and setting up of one-stop service centres at police stations.

To summarise, women's education, health statistics and labour force participation are improving. Violence against women is decreasing, although there is certain fluctuation; it is unclear to what extent this is caused by a change in reporting rates as opposed to a change in prevalence.

Table 1.3 Reported violence against women in Bangladesh from 2013 to 2017

Year	Rape	Dowry	Acid Violence
2013	3650	5716	71
2014	3689	6867	49
2015	3828	6595	46
2016	3717	5278	20
2017	3979	4125	5

Source Gender Statistics of Bangladesh, 2018, BBS

1.5.2 Disaster, Women and Vulnerability

Disaster social scientists predominantly conduct quantitative research in which they use only gender as a demographic variable (Bolin et al., 1998; Fothergill, 1998). Certain feminist researchers have attended to gender relations and how they can intensify and increase vulnerability of women during disasters (MacGregor, 2010). Despite extraordinary advances and strong women's movement for women's human rights, women's lives are dependent on education and circumstances, on power and oppression, on poverty and livelihood capacity and on their links to land, information and training (Alston, 2015). In this part of the study, researchers focus on the literature to understand the discourses of disaster and women relating to socio-cultural norms, social institutions, migration, violence, health, patriarchal norms, female-headed households, entitlements and involvement of women with disaster management processes, to clarify disproportionate vulnerability, coping and adaptation mechanisms, women's involvement with decision-making processes, involvement of women with income-generating activities and changes and improvements to ways of life, local norms and institutions.

Although disaster affects both women and men, the physical burden of coping is dependent on women (Khondker, 1996; Nasreen, 2012). Disasters increase women's domestic burden, create great hardship and undermine women's well-being because of their dependence on economic activities linked to the home. Moreover, women are more vulnerable because of the effects of climate change when they are highly dependent on local resources for their livelihoods (Alston, 2015).

Poor women suffer from lack of food, clothing and shelter, whereas young girls suffer from insecurity. Often, unemployed men stay idle or migrate somewhere else, thus leaving their household members behind. Under these circumstances, women take responsibility for caring for their houses, children and other members of the family, livestock and belongings. To ensure the survival of their families, many women in Bangladesh remain hungry to support their children and work long hours in their old age (Alston, 2015).

Because of the gendered division of labour, poor women are more exposed to natural hazards compared to men. Blaikie et al. (1994) argued that the poor often involuntarily reside in disaster-prone areas, which is reflected by Chowdhury (2001). They argued that, in Bangladesh, man dwellers are usually engaged in waged labour outside *chars* during the daytime, whereas women always stay at home and are possibly more at risk. Nasreen (2010, 2012) reports that during floods, men in rural areas lose their work while women carry the responsibilities of sustaining households. The traditional gender-specific responsibility becomes extremely difficult for women during a disaster such that their lives are often at risk, particularly while travelling during floods. Poorer women experience greater increases in work load (Bari, 1992; Nasreen, 2012). Gender is an important indicator of vulnerability, which is exacerbated during a disaster because of the loss of control over resources, production, decision-making, breakdown of education, unemployment and exposure of unsafe social conditions, in Bangladesh (Alston, 2015; Cannon, 2002; Dankelman, 2010; Khondaker, 1996; Nasreen, 2010, 2012).

Different studies (Demetriades & Esplen, 2008; MacGregor, 2010) show that women are disproportionately impacted by disaster, which is observed in the numbers of mortality, morbidity and injury, all of which are largely determined by gender relations, physical capacity, division of labour and socio-cultural norms. It is estimated that women and children are 90% of victims of the 1991 cyclone in Bangladesh (Cannon, 2002; Ikeda, 1995; Khondker; 1996). During the 1991 cyclone in Bangladesh, Ikeda (1995) and Cannon (2002) argued that gender-related roles, limited scope for migration and seclusion from public places keep women away from access to warning information on hazards that constrain women from making evacuation-related decisions. Sarawat, as cited in Roy (2011), reports that people who lose everything, leave their homestead and take shelter in makeshift houses on roads and damaged embankments; although the situation has considerably improved since 1991, people still stay on roads and embankments (Alam & Rahman, 2011; Paul, 2009). Because there is no arrangement for such a large number of people, they face shortages of drinking water, food, shelter and sanitation. Mismanagement, unplanned migration and displacement cause physical insecurity and traumatic experience, particularly for women. Often, increase in outgoing migration causes sexual violence and harassment against women in the place of destination and transition (Dalrymple et al., 2009).

Women are extremely likely to die during a disaster or be exposed to violence during a disaster. They are more constrained by their responsibilities while they are living in poverty, have no ownership of land and resources, have less control over production and income, lack of education, lack of access to institutional support and information and have fewer positions in decision-making processes (Dankelman, 2010 as cited in Alston, 2015). Disaster-affected women cite multiple forms of structural discrimination such as early marriage, trafficking, rape, sexual abuse and physical torture (Nasreen, 1995, 2010). Living in shelter centres without privacy and in close proximity to other people exposes women and girls to sexual exploitation. Sexual assault, domestic abuse, forced marriage and human trafficking increase in the society after disasters.

Losses of harvest, crops, livestock and poultry from floods in Bangladesh disproportionately impact women who rely on food processing, cattle and chickens for their earning (Baden et al., 1994). Women, who run home-based small business and involve in income-generating activities, lose their income, workplace and production tools (UNDAW, 2004). In addition to the financial loss because of disasters, these women do not have easy access to seek outside work (Kumar-Range, 2001) because of their household duties and restricted mobility. Consequently, earning and financial contribution of women declines compared to men's involvement in income-creating activities (Bradshaw, 2001). A previous study demonstrates that these women survivors cannot resist violence because of social stigma, economic dependence on husbands and fear of abandonment by their husbands (Bradshaw, 2001). When men become unemployed because of a disaster they release their anger on women, which includes emotional, physical, sexual and structural violence. Furthermore, during a disaster, married, unmarried and disabled women are harassed, and young girls in the 13–18 age group are the worst victims of teasing and unwanted touching; moreover, they are targeted by males/boys in the locality (Dobson, 1994; Nasreen, 2010; Sayeed, 2010).

Women's vulnerability is primarily dependent on deep-rooted and regulated social practices and structural inequalities (Alston, 2015). There are ranges of customs such as dress code (*saree*), the *purdha* system access to resources, and unequal rights on properties that hinder the ability of women in Bangladesh to effectively respond and limit their capacity (Alston, 2015). When the assets of women are depleted and income-earning options become inferior, the bargaining capacities of women in the household become weak. Consequently, women become less mobile, leaving men in crisis in a stronger "fall-back position" (Agarwal as cited in Enarson, 2000). The dependency of women on men's wages creates a condition of vulnerability, particularly as women lack legal entitlements to share their husband's wages (Dreze & Sen, 1989; Kabeer, 1990).

During a disaster, vulnerable and disadvantaged groups such as women and widows, who are the head of the household, suffer the most and are discriminated against when they access resources and assistance in Bangladesh. During the post-disaster period (relief and rehabilitation period), multiple widows were not entitled to claim property and resources that are legally owned by their husbands (Oxfam, 2005; Rosa, 2005) and a male-headed household is considered as a better model. Female-headed households or those abandoned by their husbands often fail to receive attention from males in authority (Bolin et al., 1998; Chant, 2008). Because these women are abandoned, divorced and separated from their husbands, it underlines that women heads have less or no ties with ex-partners' relatives with their own families and with communities. Consequently, they are not respected, lose their dignity in the family and society and are ignored and deprived of material resources.

1.5.3 Disaster, Women, Coping and Adaptive Mechanisms

Although media show over and over distressed and crying women with their babies and highlight women as sufferers (Bolin et al., 1994), women's individual adaptive capacity and initiatives turn out to be central for their family nourishment and food security to overcome the crisis. In all relief lines, women stand first and they are collecting wood and bamboo to reconstruct the residences. As is customary, they deal with the children and lack of food. Women who are responsible for fetching water might become weaker from water-borne diseases, which might be more prevalent among women, who are nutritionally deprived (Cannon, 2002). Women are both victims and active agents to cope with a disaster. Paolisso and Amanda (2002) demonstrate that the women survivors of Hurricane Mitch assessed the losses of their agricultural products and coped with the disaster based on their knowledge and social networks.

However, the sources of drinking water are becoming a major problem for people of southwest Bangladesh with increase in climate variability and disasters. Nevertheless, it is once again the duty of women only, irrespective of their physical condition, to supply drinking water for their families. Because water sources in the surrounding area are affected by pollution and high salinity, women travel long distances to collect drinking water when both ground and surface waters are made saline (Khatoon, 1995).

Despite heavy burdens, challenges, traditional division of labour, social perception and norms, women confirm their capacity to face extremely harsh circumstances and demonstrate their strength and resilience to protect their lives and resources in Bangladesh (Gokhale, 2008; Khan, 1995; Nasreen, 1995, 2012). Fernando and Fernando (1997) argue that women manage crisis situations and demonstrate high degrees of resilience for victimising relationships. The role, responsibility and sensible contribution of women in Bangladesh make the society sustainable and women are the backbone of rural economy.

Women have important contributions in creating a doorway to natural, financial, physical and social capital and make their livelihood sustainable (UNIFEM, 1998). For example, during agricultural activities, women contribute for all phases such as seed sowing, harvesting and processing of crops. Women participate in farming activities and small farmland and a homestead area are being used more effectively by women using their own knowledge (ADB, 2004). However, at the times of crisis, it is the women who adjust their cooking and consumption patterns by sacrificing their own meals for male family members. Women's work in agriculture is extremely common, seen as an extension of their domestic responsibilities rather than a separate economic activity.

In rural areas, storage of food, collecting relief, water and fuel and caring for children, elderly people and livestock are more predominantly taken care of by women (Gokhale, 2008; Khatoon, 1995; Khondker, 1996; Nasreen, 2012). The socially dispersed role of women as caregivers and nurturers unsurprisingly extends to risk management in the continuation of secured livelihoods and life support systems during a disaster (Ariyabandu, 2003). The aid and assistance from different GO and NGOs are not sufficient to meet the family's requirements. Women make different handicrafts and mats (Azad, 2013) to sell in local markets along with livestock and poultry rearing, backyard gardening and planting vegetables. Women work and contribute as volunteers, nurses, day labourers, paid workers and family members. They take economic responsibility for the family during a crisis. Women's role in preparing and storing food items for the family for days at the time of disaster are the source of energy and nutrition. For example, the traditional measures of storing belongings by the women such as clothing and bedding, which they can be easily removed at the time of a disaster (Ariyabandu, 2003). In the Bangladeshi coastal area, women are the members of community-based committees and participate in innovative livelihood programmes initiated by GO and other organisations to build resilience. Mitchell et al. (2007) demonstrate that the Bangladeshi women of flood-prone areas store seeds in high places in the house using their traditional knowledge. They may easily take with them to safer places during a flood.

The coping and adaptive mechanisms of women during a disaster in the Bangladeshi coastal area range from safety measures at home to decision-making at household, local, regional and national levels. Women are being included in disaster management committees for different levels of disasters and addressing safety measures at home, participating in community disaster planning, participating in awareness programmes, learning construction skills, caring for vulnerable people, providing emotional support and assisting in relief distribution. During a crisis, women take responsibility for looking after family, ensuring food, medicine and household livelihoods and demonstrating a proactive nature (Oxfam, 2011). While facing a disaster, women gain knowledge and skill using their own perspective and attitude and learning strategies to cope and adapt. Although the risk reduction strategies for women differ from formal skills provided by GO and NGOs, agents, policies and training, women play an important role to minimise their own and family losses (Oxfam, 2011). Through everyday practices and activities, women exercise their coping and adaptive strategies.

Chowdhury (2001) shows that the indigenous, traditional knowledge and innovative technology of women plays an important role in managing environment and protecting lives and livelihoods of the family. For example, composting kitchen waste primarily by women provides a soil-enriching fertiliser, by planting trees around houses women protect residence from strong winds, and by selecting fast-growing seedlings made soils more stable (Chowdhury, 2001) in the quickest possible time. Ariyabandu and Nivaran (as cited in Enarson, 2000) report that with the onset of floods that, when men are out of the locality, it is women who take cattle or utensils from their house to safe places. If required, they bind trees with ropes and climb on that with children and elders. By the time this eventuality takes place, hardly any men are left in villages and women have to manage everything.

1.6 GO, Institutionalisation of Disaster Management and Women

Historically, NGOs in Bangladesh, as an agent of social change, have been recognised and acknowledged worldwide. However, the contribution of NGOs solely after the birth of Bangladesh was not sufficient to reduce the increasing damage and losses, including life and property, from disasters. Despite certain strengths, such as long experience of coping with disasters, the people's resilience, support from NGOs and donors, the GO has been unable to deliver services during a disaster because they suffer from a host of policy and institutional weakness (Khan & Rahman, 2007). The Bangladesh GO has adopted various coordinated programmes and policies for disaster management to protect from intensifying disasters in future and for building an institutional collective disaster management by involving GO, local communities, NGOs, the private sector, academia and donor communities. The vision of building institutional and collective disaster management of the government is to reduce the vulnerability of people, particularly the poor, from the effects of natural, environmental and human-induced disasters to a manageable and acceptable humanitarian level.

The vital mission of the MoDMR of Bangladesh GO introduced a paradigm shift in disaster management from a conventional relief-based approach to a comprehensive risk reduction culture for ensuring the resilience of communities to hazards. For this purpose, the Bangladesh GO has effectively created a well-defined disaster management institutional mechanism via SOD issued in January 1997. In addition to all agencies, the MoDMR is responsible for coordinating national disaster management efforts. Under SOD, many inter-related committees, both at national and sub-national level, have been formed to ensure effective planning and coordination of disaster risk mitigation and emergency response management at all levels, which has then been adopted via NPDM 2008–2015 and 2010–2015. Many inter-related institutions and institutional networks have been developed from national to local levels by forming disaster management committees.

Networks developed from national to grassroots level and institutionalisation of disaster management committees strengthen the capacity of the disaster management system. The development of networks in disaster management by the Bangladesh GO is a holistic approach on which emphasis has been given for working as a team with all relevant agencies, including NGOs, stakeholders, private sectors, donors, local people and local institutions. As a governmental institution, NPDM deals with initiatives after the creation of DMB in 1993 with the aim of addressing important issues, including risk reduction, capacity-building, climate change adaptation, livelihood security, gender mainstreaming, community empowerment and response and recovery. DMB is the coordination body that aims to create public awareness in terms of preparedness and coordinates all activities from national to grassroots levels, including NGOs (O'Donnell & Wodon, 2015). NPDM is committed to SOD, the active instrument of CDM of the country introduced by MoDMR and DMB. Bangladesh introduced its CDMP with financial and initial logistical support from UNDP for mainstreaming a disaster management programme in development plans before introducing the HFA Plan, MDG and UNFC on climate change (GoB, 2010). For mainstreaming gender and disaster management programmes in the development process, CDMP organised knowledge and skill development training for academics, students, civil society members, NGO activists and development workers around Bangladesh.

1.6.1 GO, Disaster and Climate Change Policies in Bangladesh and Women

Bangladesh, as a newly independent (1971) country, actively considered strong measures in support of gender equality and women's empowerment from the beginning. Moreover, to ensure its continuity, Bangladesh actively participated in World

Conferences on Women in Mexico in 1975, followed by a fourth conference in Beijing in 1995 that focused on human rights for women and girls. Unfortunately, measures taken by the new GO in a newly independent country were ravaged by the killing in 1975 of the leader of the independence movement and president of the country, Bangabhandu Sheikh Mujib, using a military coup that ravaged Bangladesh in all sectors. Because of assassination, martial law, led by General Zia, took over power of the country and that continued until 1990, led by General Ershad. This slowed down the initiatives for gender equality and women's empowerment. However, the slow incorporation of religion in the state affairs by the military rulers, and the adoption of Islam as the official state religion, changed the national commitment to gender equality (Alston, 2015), although the 1973 constitution guaranteed equality and non-discrimination of sex, religion and ethnicity in the public and private domain.

After three decades of independence, particularly after 1990 when democratic rule returned after the 1975 military coup, changes occurred and upheld the status of women to certain extent; however, the achievement is not at a satisfactory level. Efforts towards bringing women into the development process in Bangladesh are based on a wide array of international commitments, including CEDAW (1979), Beijing Conference (1995), UN Decade of Women (1976–85) and MDGs (2000–2015). After the guidelines and declarations of the CEDAW, Bangladesh GO involved women with the different development programmes for the betterment of women. For example, policies, strategies and action plans in terms of climate change and women's empowerment include the BCCSAP, (2009a), NAPA, (2009b), ccGAP: Bangladesh, (2013), Climate Change and Disaster Management 2015, Vision-2021 (a strategic plan/road map for Bangladesh till 2021) and the Sixth (2011–2015) and Seventh (2016–2021) FYPs. The policies matter because they represent the country's intention to empower and respond to requirements of the often invisible "other half" of the population (GoB, 2013). If women had the same access to resources as male farmers, farm yields could increase by 20%–30% and the number of hungry people could be reduced by 12%–17% (FAO, 2011 cited in GoB, 2013). Nasreen (2008) argues that MDG goals and targets, i.e., climate change and disaster-related issues, are set to ensure the rights of women and men, including environmental sustainability.

Furthermore, the Bangladeshi government has adopted various coordinated programmes and policies for disaster management to protect from intensifying disasters in future and for building an institutional collective disaster management by involving GO, local communities, NGOs, the private sector, academia and donor communities. The vision of building institutional and collective disaster management of the government is to reduce the vulnerability of people, particularly the poor, from the effects of natural, environmental and human-induced disasters to a manageable and acceptable humanitarian level. For this purpose, the Bangladeshi government has effectively created a well-defined disaster management institutional mechanism via SOD issued in January 1997. The Ministry of Disaster Management and Relief (MoDMR) is responsible for coordinating national disaster management efforts along with all agencies. Under SOD, a number of inter-related committees, both at national and sub-national level, have been formed to ensure the effective planning and coordination of disaster risk mitigation and emergency response management at all levels, which has been adopted via the NPDM 2008–2015 and 2010–2015. A number of inter-related institutions and institutional networks have been developed from national to local levels by forming disaster management committees.

Networks developed from national to grassroots level and institutionalisation of disaster management committees strengthen the capacity of the disaster management system. Network development in disaster management by the Bangladesh government is a holistic approach in which emphasis has been given for working as a team with all relevant agencies, including NGOs, stakeholders, private sector, donors, local people and local institutions. As a governmental institution, NPDM deals with initiatives after the creation of the DMB in 1993 with the aim of addressing the important issues, including risk reduction, capacity-building, climate change adaptation, livelihood security, gender mainstreaming, community empowerment, response and recovery. DMB, the coordination body, aims to create public awareness regarding preparedness and coordinate all activities from national to grassroots levels, including NGOs (O'Donnell & Wodon, 2015). NPDM is committed to SOD, the active instrument of CDM of the country introduced by Ministry of Disaster Management and Relief and DMB. Bangladesh introduced its CDMP with financial and initial logistical support from UNDP for mainstreaming a disaster management programme in development plans before introducing HFA Plan, MDG and UNFC on climate change (GoB, 2010). For mainstreaming gender in disaster management programmes and in the development process, CDMP organised knowledge and skill development training for academics, students, civil society members, NGO activists and development workers all around Bangladesh.

Recently, focusing primarily on building adaptation capacity, the Bangladesh GO established a National Climate Change Fund (from national and international sources including UN) with an initial capitalisation of $45 million, which was later increased to $100 million (Ahmed et al., 2015). The Bangladesh GO placed importance on developing a pro-poor climate change strategy for achieving economic and social well-being for the well-being of all people where the poorest and most vulnerable groups, including women and children, receive special attention. In the Climate Change Action Plan [a 10-year programme (2009–2018)] the requirements of poor and vulnerable people, including women and children, have been prioritised in all activities of the country to meet the climate change-related challenges (GoB, 2009). Being the primary and direct victim of climatic events, women play an important role, particularly when it comes to decision-making and building community resilience in the face of climate change and the meaningful participation of women becomes important (GoB, 2013).

GoB (2016) suggests that considerable efforts have been made to understand gender-distinguished impacts and adaptive capacity of climate change on both women and men. In particular, the adaptive capacity of women has been recognised as important actors for implementing ccGAP: Bangladesh. Moreover, the 7th FYP recommended the development of institutional adaptation addressing people-oriented community-based adaptation at a microscale. Ahmed et al. (2015) explained the importance of gender mainstreaming in environmental efforts and poverty eradication in a wide range of global agreements and conventions such as the Rio Convention. By signing agreements such as UNCBD and UNFCCC and confirming such agreements and conventions, the Bangladesh GO has equipped their policies such as FYPs with a specific focus on gender equality and women's empowerment. In response to the requirements of women, and to focus women in the context of climate change and to respond to extreme climatic events, in ccGAP: Bangladesh policies, strategies and interventions have been included and gender issues have been integrated under four primary pillars, namely, food security, social protection and health, comprehensive disaster management and infrastructure and mitigation of low carbon development.

Climatic events such as cyclone, droughts, excessive rainfall, flood, salinity and erosion have already created food insecurity globally. Sustainable food security is deteriorating on a daily basis and intensified climate change will make it harder. The poor and other vulnerable groups, i.e., women and children, are possibly at high risk of food security because of climate change (GoB, 2013). Under food security, social protection and health pillars, gender and climatic concerns have been integrated in policies on agriculture, environment and water resources for women. Furthermore, it has focused different social safety nets for women such as capacity development, thus ensuring access to technologies and alternative technologies (i.e., bio-fertiliser and climate resilient crops). GoB (2013) integrated climate change gender health issues to support livelihood for women to ensure access for safe water and better social protection for women in times of disaster.

Frequent and extreme climatic events have principally bad effects on key determinants on public health, including clean air, water, fuel wood and food. Polluted air and water, lack of safe water and sanitation create a favourable atmosphere for spreading infectious and climate-sensitive diseases such as food-, water- and animal-borne diseases, and GOs put effort in building adaptation to livestock and the health sector, including water and sanitation in climate-vulnerable areas, and livelihood protection for socio-economically vulnerable groups, including women (GoB, 2009, 2013). Ensuring immediate access to food, income-generating activities, salinity resilient seeds and training enables poor households, women and female-headed households to sustainably deal with the impacts of climate change and protect their livelihood strategies (GoB, 2011a, 2013). In terms of the gender mainstream, to increase the resilience of vulnerable groups, particularly women and children, the 7th FYP (2016–2020) has recommended the development adaptation and livelihood diversification at a community level. Note that additional argument exists for ensuring drinking water and sanitation in the locality because polluted water and unsafe sanitation create risks for climate change.

For women who are landless and who have small farm and are marginalised, the norms are traditionally much more relaxed and women from these types of households are relatively more engaged in agriculture and even in field activities (GoB, 2013). Moreover, in recent years, poverty, male migration and women's empowerment and mainstreaming of women in the development process have led to "feminisation" of the agricultural labour force in Bangladesh and women form 45.6% of total farming population (GoB, 2008). In the absence of male members, the roles of women have changed from unpaid family workers to paid worker sometimes as farm managers (Jim & Hossain, 2011). Women are constrained by social, cultural and economic factors. To overcome these limitations, ccGAP: Bangladesh (2013) made the suggestion of providing necessary tools, fertilisers, loans and farmers ID card to women farmers to ensure food security (GoB, 2013).

Similarly, National Agriculture Policy (2010) focuses on mainstreaming women in agriculture because the involvement of women with production and marketing ensures women's access to credit and micro-credit. In moving the policy forward to involve women, Naved (2011) reports landless, small and marginalised farm households have relatively flexible gender norms and these groups are responsive to interventions such as gender in agriculture and women are dealing with them more efficiently. Women require to be linked to market and would be enabled to access and control their own income. For food and health security, ccGAP: Bangladesh (2013) ensures women's access to land tenureship/leasing for cultivation by motivating land and water body owners to lease land/water to women. The above-mentioned policies emphasised the management of community clinics and union health centres to make them function more effectively for women at a grassroots level as well as adolescent girls.

Under the CDM pillar programmes, such as the development of a gender-responsive disaster management policy, ensuring women's participation from central to local UDMC, additional financial allocation for women, capacity development activities for women and health care for women are considered as a prerequisite for sustainable development. Both GoB (2009b) and GoB (2009a) focused on strengthening the capacity of GOs, civil society and communities to manage climatic disasters. This ensures that appropriate gender-neutral policies, laws and regulations are in specific places to be correctly practised. This suggests strengthening community-based adaptation programmes that are already in the disaster-prone parts of the country where vulnerable groups and women are prioritised. However, the 6th FYP (2011–2015) shows that the Bangladesh GO supports communities and rural people all the way to strengthen resilience and adaptive capacity to face climatic challenges (GoB, 2013).

In terms of sustainable hazard and disaster management, GoB (2016) suggested for strengthening capacity of GO by involving civil society partners, volunteer organisations and community people. The realisation that women face the challenges of risks associated with disasters differently to men has been widely accepted across the world; however, in Bangladesh, it has only recently been recognised. In a short time period (since NGOs became involved in the 1990s and policies changed around 1997), women have become more resilient (Nasreen, 2012) and demonstrated ingenuity using indigenous/local knowledge and acted as community mobilisers in disaster response (WEDO, 2008) despite challenges. To speed up the greater involvement of women, the GoB prioritised the involvement of women with CDM at a national policy level (GoB, 2013). For a gender-responsive policy in the context of disaster management, climate change and sustainable development decision-makers and policy planners are motivated for farming/agricultural work gender-sensitive policies and to introduce gender-budgeting across all GO agencies.

Many studies show that women tend to be more concerned about climate change and they can put in effort to reduce GHGs more than men (GoB, 2013). For example, women are culturally more habituated to plant trees around households, which reduce carbon emissions and are an alternative source of income and household fuel. In Bangladesh, the idea of "social forestry" in the 1980s, with a financial loan from ADB created an alternative source of income for women and local poor people and to solve fuel crisis in rural areas. Karlsson (2007) argue, in developing countries (including Bangladesh), fuel energy for cooking, heating and lighting is obtained from wood, charcoal or agricultural waste, which is customarily a woman's work. Women are considered during policy development because they provide family and community care, and they are important consumers of energy in rural areas. Women in communities can work together in preserving a healthy community and both men and women can benefit from the positive aspects of climate change policy. For example, opportunity for women to work in renewable energy and waste management in rural areas will make women financially solvent and they can do this staying at home (GoB, 2013, 2016).

Ahmed et al. (2015) reported that the promotion of renewable energies that help avoid GHG emissions could provide an approach for promoting women's economic participation in climate action. Women in communities can collaborate in maintaining a healthy community environment while economically benefitting by generating additional needed income. One report (Gender cc, 2016: www.gendercc.net) mentioned that women are more willing than men to change their behaviour to save energy and purchase low carbon-emitting products. Women reject risky technologies and prefer to rely on lifestyle changes rather than only on technological progress. Policy planners and decision-makers incorporated gender issues in climate policy and agreeing that women deserve more attention in policy. Gender-related considerations are addressed and women's participation as a group is incorporated in the policy to

- promote gender-responsive programmes to reduce GHGs at a household level.
- improve women's knowledge on climate change mitigation issues raising awareness using local media, e.g., poster, miking and drama.
- mainstream gender consideration in coastal and social forestry programmes to expand coastal green belts, social forestry and plantations on river and coastal embankments via women's participation.

Climate change-related policies initiated working collectively to face disaster and to facilitate the creation of gender equality policies and their translation in practice. At this point, the state as a key institution began to shape and control gender relations and to create novel language and frameworks that accept women's rights as a fundamental principle (Alston, 2015) for development. GO confirmed that not enough had been done in the past for women and they are doing more now compared to the 1990s, although their journey for women's empowerment and gender equality has been disrupted by political unrest, lack of cooperation and trust among the political parties, religious fundamentalism/extremism, the boycott of the last national election by a major political party, blaming of civil society and blaming of GO by civil society. However, the efforts of government; the policies; and the generation of support, dissemination and application of information have improved the knowledge management systems that improve capacities (Ayers et al., 2014) and speed up changes that have already happened in the lives of women. This is manifested in all sectors including *purdha* practice, involvement of women with earning and outdoor activities, direct participation of women in politics and disaster management system which is discussed in Chapter 6.

1.6.2 GO and Mainstreaming Women in Disaster Management

The involvement of women in awareness-building activities and participation with decision-making, policy-making processes, community empowerment, capacity-building and responses relating to disasters improve women's skills. Both women and children are given special care to take in safe places/shelter when there are disaster warnings that resulted in reducing risk to women and children and social and their economic power has been improved. For example, from FYP 2011–2012 to 2014–2015, 61,654 women in total have been trained for managing disaster, including 10,000 women from CPP volunteers and 8,000 from fire service and civil defence. Moreover, 9540 women members of the disaster management committee at district, upazila and union levels have been trained. The number of trained women in disaster management, volunteers, fire service and civil defence, is in sufficient compared to the millions of people in the coastal area. Despite good intentions, the number cannot be sufficiently increased because of the limitation of resources but it is a positive move and start.

1.6.2.1 Community Approach and Women in Disaster Management

The MoDMR procured different search and rescue vehicles/equipment to combat disasters, thus playing an important role in saving lives, particularly women and children, and property during disasters. Despite exemplary progress and implementation of different social safety net programmes, the construction and maintenance of flood and cyclone centres as well as the process of mainstreaming women in disaster management is not linear. Because these projects, policies and plans for mainstreaming of women are not distinct because of different challenges, e.g., inadequate honest will of coordination between GO organisations and NGOs, among local representatives, beneficiaries and community leaders, limited capacity knowledge of GO officials about locality, nepotism, political bias (details in Chapter 5) has been noticed by respondents. These constraints led to delays in the development agenda and led to unpredictable progress in mainstreaming women in the disaster management process. However, perceived information and evidence, bias evidence based on political drivers for decision-making leads the policies to uncertainty, delay and slow implementation (Ayers, 2014).

In addition to the local initiatives by GO for women's leadership in disaster management in 2004, Bangladesh GO, with the support of UNDP, launched the CDMP to reform the disaster management approach by expanding its focus, including policy advisory services, capacity development, risk mitigation and effective humanitarian response. Moreover, the GO programme and institutions are different international development partners such as Red Crescent Society, Oxfam, Care Bangladesh, Action Aid Bangladesh, Islamic Relief and Swiss Contact, thus providing support to the Bangladesh GO and local development partners and NGOs to ensure a humanitarian response for all.

In the last decade, the disaster management sector in Bangladesh, based on its experiences, has introduced a community-based disaster management approach that aims to build resilience and leadership of women to reduce vulnerability (MoFDM, 2005). An effective DRM process must be participatory, decentralised and local experience-based and bring a larger section of population into the public decision-making processes, which is important in the context of disaster risk mitigation with greater efficiency and equity in developing resources. To develop effective leadership of women and a bottom-up approach, the GO has ensured the inclusion and participation of women, a decentralisation that includes representation of community people and direct involvement of women at all committees levels such as UDMC, PDMC and TDMC by the Local Government Ordinance, 1983 and SOD, 1997. As members of TDMC, PDMC and UDMC, female members attend to different meetings and workshops related to community vulnerability, community risk assessment process and risk reduction action plan organised by both GO and NGO. Participation enriches their knowledge to mobilise other people in the community during a disaster and utilise the resources for implementing the risk reduction action plan (GoB, 2013).

Messer (2003) confirms that disasters/hazard risks normally do not affect Bangladesh and the impact of hazards/disasters vary from one small region to another. Therefore, it is important to develop community-based leadership at a local level using local knowledge. Community-based leadership based on local knowledge creates effective prevention measures to adapt to local hazards and vulnerabilities. GO decision-making processes along with the citizens of community are believed to increase public sector accountability. Crook and Manor (as cited in Messer (2003)) confirmed that when GO becomes closer to community people, it increases efficiency and creativity regarding the use of resource of local communities. Therefore, effective measures contribute to strengthen the common people-oriented participatory culture, which has been reported in the coastal area. For example the increasing number of women in democratic practice such as direct participation and representation of women in election is a positive indication although it is not remarkable but the increase in democratic practice and representation of women. However, although women are getting attention in the disaster management efforts by different actors, considerable efforts are not yet given in a systematic manner (Nasreen, 2012) and the government is making efforts to implement the policy and not allowing sufficient time to evaluate how it locally works.

1.6.2.2 GO, Disaster Risk Reduction and Women

Women in the Bangladeshi coastal region are evidently more vulnerable to disasters compared to their counterparts. Recognising their different roles, capacities, requirements, and considering them in risk reduction strategies, is important for reducing women's vulnerability to disaster. For this purpose, it is important to simultaneously recognise the varying requirements and capacities of women and integrate them by mainstreaming women in disaster risk reduction (Ginige et al., 2014). The mainstreaming of women in risk reduction, adaptation of climate change and disaster involves the integration of information, policies, stakeholders, beneficiaries, vulnerable community and measures; moreover, integration is considered as a tool for sustainable, effective and efficient use of resources (Ayers et al., 2014). To mainstream gender in risk reduction of disasters, Bangladesh has approached risk reduction and adaptation of climate change-specific plans, programmes and institutions that address various aspects of vulnerability and climate risks for development programmes and policies (Ayers et al., 2014).

In Bangladesh, the mainstreaming of gender in disaster risk reduction is seen in much the same light as poverty reduction. The high mainstreaming of women in disaster risk reduction programmes results in increase in the efficiency to deal with disasters in the most efficient manner (Bradshaw, 2013). Moreover, the application and incorporation of gender in disaster risk reduction strategies promotes gender equality, increases community awareness, ensures commitment of GO, stimulates inter-disciplinary and inter-sectoral partnerships and increases the scientific knowledge of community people. It allows the community to justify that women are targeted not because they require help rather they are seen as being better at risk reduction, as well as women being better able to distribute resources to other vulnerable people. Resources given to women after a disaster are used more efficiently and in a better way (Bradshaw, 2010; Ginige et. al., 2014). Gender mainstreaming by GO is an outcome of many top-down and bottom-up interventions that have been summarised below (Fig. 1.1).

1.7 NGOs: The Alternative Platform and Women Empowerment

In the post-independence era in Bangladesh, considerable improvement of women's well-being and empowerment of women are seen in different sectors, e.g., via education, health care, participation in the labour force and politics. Empowerment is a

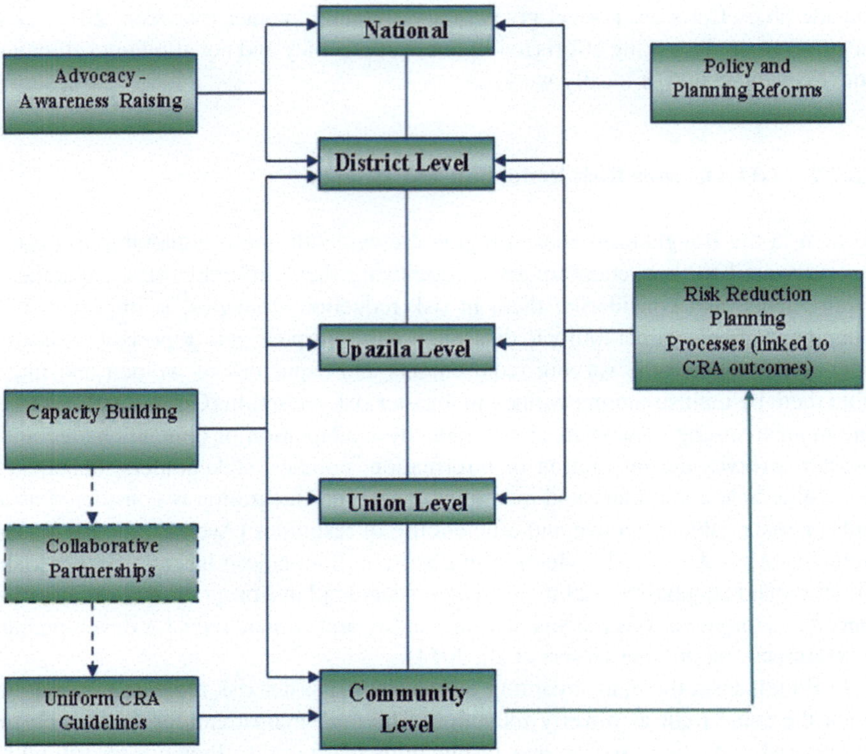

Fig. 1.1 Mainstreaming Strategies for Risk Reduction (*Source* National plan for disaster management 2010–2025)

process where individuals, by gaining control over their lives, attempt to reduce powerlessness and dependency (Lord & Hutchison, 1997 as cited in Panday, 2016). Social mobilisation is creating awareness and increasing women's rights, including political, social and economic rights; more importantly, it is empowering women by building agency (a person's ability to define goals and to act upon them; in other words, a person's ability to make decisions). This motivates women in getting out of the kitchen to attend meetings (Panday, 2016) and earning and bringing out their demand in public space.

After declaring the UN International Year of Women and the subsequent UN Decade of Women (1975–1985), the Bangladesh GO, national and international NGOs, donors and civil society undertook several programmes for the progression of women in Bangladesh. These milestones assisted the development of the women's movement in Bangladesh and groups set about documenting the status of women, recognition of women's work and women's legal status (Alston, 2015). The women's movement has played an important role in increasing mass consciousness in terms of women's concerns and increasing women's participation in every stage to achieve equality (GoB, 2013).

In addition to the state, the role of NGOs remains important for making people realise what women must have to ensure their active participation for all development processes. Addressing local requirements such as poverty eradication, employment generation, public health and basic education has been increased and NGOs have struggled to open up avenues for the participation of women. Consequently, women occupy a significant position within national political, economic, socio-cultural and development discourse through their contribution as directly elected representatives at both national and local level, garment workers, community mobilisers and going abroad as skilled and unskilled workers.

Development ideas, empowerment of women and changes in women's lives were primarily influenced by donor agencies in the 1970s and 1980s and were later taken over by national NGOs, and women's organisations such as BMP, NP and WFW, and this still continues (Nazneen et al., 2011). These women's organisations were working along with GO officials who were being exposed to international forums discussing women's rights that came to the forefront because of women's movement, which continued to expand, starting with urban areas and moving out to villages via NGOs under the name of women's empowerment programmes (Alston, 2015). NGOs attempt to overcome deeply rooted social and religious taboos, limited finances, poor governance, and corruption, all of which are slowing down change-related initiatives (Panday, 2016).

Note that ~ 80% of villages in Bangladesh have an NGO programme. This is fuelled as alternative service provider in the 1980s with the aim of providing basic services, including economic, social, health care, non-formal education, legal aid, social safety net programmes and sanitation. Initially, the specific focus of NGOs was to promote poor rural women as development service users, economic agents and development workers (Goetz, 2007; Nazneen et al., 2011; World Bank, 2006), and they successfully reduced gender-based discrimination, inequality and deprivation. NGOs emerged as complementary avenues for rural poor women's empowerment (Nazneen & Sultan, 2009; Nazneen et al., 2011). NGOs and women's organisations contributed to integrating/mainstreaming women by creating opportunities for funding various gender projects in Bangladesh. By expanding the NGO sector and their activities, the social, cultural and political context in Bangladesh is upheld (Nazneen et al., 2011).

Because of the location, operational area, nature of activities and target/beneficiary group of NGOs, they were able to popularise their mission and vision for building networks and agency for empowering people at a grassroots level, particularly women. Foreign donor agencies found it attractive to assist NGOs. Because NGOs as grassroots organisations were independent from the state, it was easier than directly dealing with states (Panday, 2016). Without developing women's personal agency, it is difficult to ensure women's empowerment and entitlement to resources (Panday, 2016; Sen 1985). International NGOs picked up the point of empowering women and played an important role in supporting human rights and gender inequality in Bangladesh via aid funding, building partnerships with local organisations and firming ideological commitment to gender equality (Alston, 2015).

Economic development shows better progress when women become involved in and take a lead in development-related projects (Panday, 2016). In Bangladesh, women started to exemplify the success of economic empowerment at an individual level by implementing the micro-finance model (Nazneen et al., 2011) that was introduced in mid-1990s by NGOs such as Grameen Bank and BRAC. Bangladeshi women have experienced empowerment and mainstreaming from collective aspects after the mid-1990s (micro-finance and ready-made garments) that addressed women's political emancipation, empowerment, advancement and status. The empowerment of women was emphasised when NGOs, donors, feminists and activists focused on women as an inclusive part of society and it conceptualises that changes for women are changes for society. Leading NGOs such as BRAC, Proshika, Grameen Bank, BMP, NP, WFW perceive women's empowerment as requiring a holistic approach. They believe the empowerment of marginalised, poor and underprivileged, subaltern groups for extensive, intensive and participatory process of sustainable development will come through women empowerment (Nazneen et al., 2011).

In terms of the empowerment of women, they (BRAC, Proshika, Grameen Bank, BMP, NP, WFW) argue that they can deliver welfare services more efficiently to the poor because of their principles of participation, empowerment, accountability, participatory approach towards development at a grassroots level and close proximity to poor, marginalised and underprivileged people (Kabeer et. al., 2012; Panday, 2016). In addition to different formal women organisations in Bangladesh, with the direct and indirect patronage of NGOs, multiple informal *samity, sangstha* and *shangathan* operate at a rural level. They are more popularly known as *mahila samity* and they provide training, distribute relief and help women to be engaged with multiple cultural and social activities that ultimately facilitate their empowerment. Households and communities are recognised as spots of women's empowerment that start by ensuring women participate in economic activities. Involvement with income-generating/productive activities makes women confident and respectable to family and society and makes pathways for women to be part of the decision-making process for both the household and community. Nazneen et al. (2011) argued that the destitute rural woman, i.e., the worst affected, can play a decisive role as agents of change if she is economically empowered, and NGOs believe in empowerment as a core strategy for social change.

The political empowerment of women is important to changing women's position and increasing their bargaining power for all decision-making spheres from policy-making to school committees (Nazneen et al., 2011). However, the political empowerment of women is possible only if more women can participate. If women's direct participation in the political process is an equality issue, it is expected to enhance women's economic and organisational capacities, self-confidence and ability to obtain a greater share of the political system. Women's under-representation at all levels of governance has generated demands by women's organisations and NGOs for greater political participation and representation of women (Panday, 2016) to bring women (as leaders/representatives/decision-makers) closer to the mass people and make it easier for women to enter the public domain (Nazneen et al. 2011).

To address the importance and movement for the greater participation of women in politics, quotas for women both at national (50 reserved seats out of 350 since 2004) and local (three reserved seats for women out of 12 since 1997) level have been introduced in Bangladesh (Panday, 2016). Although the quota for women in politics is not at all satisfactory, it is positive that the participation of women in politics is increasing and their voices are increasing. Accordingly, different NGOs and different women's organisations have been advocating in favour of women's involvement in the development, decision-making and policies of politics since the 1980s (Nazneen & Sultan, 2012; Nazneen et al. 2011; Panday 2016).

The involvement of women with the development, politics and decision-making process have speeded up on a large scale because > 80% of members of NGOs in Bangladesh are women (Panday, 2016). Non-state actors, including development-oriented NGOs, women's NGOs and women's organisations such as village-based *samities*, have helped link women with different organisations in different societal sectors, including policy spaces, better organisational capacity and increased outreach (Nazneen & Sultan, 2012). At a grassroots level, women members of NGOs, women's organisations and village-based *samities* play an important role in mobilising women to be included in decision-making processes, politics and development processes because if women cannot become effective in politics, they will not enjoy many benefits to which they are entitled (Panday, 2016). Women members of NGOs and *samities* play roles to build capacity, agency of the poor and marginalised sections of society, particularly for women, such that they can raise their voices for rights. NGOs are concerned with the extent to which women are able to challenge constraints imposed on their lives and consolidating democratic practices by encouraging more citizens' participation and ensuring representation of the interests of marginalised groups at grassroots level (Panday, 2016).

In terms of disaster and vulnerabilities, the affected communities show that responses to disaster with or without facilitation of GO agencies are deployed at a grassroots level. Community people, particularly women, depend on community knowledge to reduce vulnerability, which has been acknowledged and encouraged by many across the world (Ahmed et al., 2015). Community-based knowledge/local knowledge for resilience, gender sensitivity and gender-related services in disaster management and climate change has been addressed and gained attention because of the continuous effort by donors, international NGOs and national NGOs (Ahmed et al., 2015). For mainstreaming women with disaster management processes and uplifting women's experiences as a response to disaster, NGOs are taking initiatives to bring women in the male domain, e.g., marketing of services. Women are exchanging views with family members and neighbours, planting more trees than before and taking shelter in safe places earlier (GoB, 2016). Most programmes of NGOs in Bangladesh at a micro-level community, including climate change programmes, are inclusive and gender-focused with a strong gender component to improve the adaptive capacity of individuals and households (Neelormi, 2010).

It is easier for NGOs to create transformation in a wide range and collective action because their activities and programmes are community-based and start from personal experience (Nazneen et al., 2011). Transformation in social and gender relations breaks the earlier tradition (such as women staying at home, less/no representation of women in politics and decision-making processes, and no/limited access to public domain). Nazneen et al. (2011) reported that NGOs remain a pioneer in their direct role in individual and collective empowerment of the poor, particularly women. In addition to these aims, NGOs focused on certain non-traditional areas, including legal aid, trafficking, funding for women's rights, organisation for networks and collective mobilisation with the aim of creating an alternative platform for women's movement for human rights. They both have ensured their activities change in the discourses of Bangladeshi women's empowerment; more importantly, NGOs and women's organisations have created pressure after 1980s on the state to address the discourse of women's empowerment. NGOs as "watchdog" have institutionalised, strengthened and created novel avenues for women in the community to express themselves and created possibilities for additional citizens' participation (Panday, 2016), which has performed much more to move beyond an isolated approach for women's empowerment. The unequal distribution of resources and influence women in the community have limited the approaches of influencing policies and practices that disadvantaged women. However, women are more mobilised and more responsive in requirement and crisis where NGOs played the role of a pioneer.

1.8 Theoretical Framework and Its Relevance

In this study, multiple theoretical frameworks have been used to frame disaster, women and vulnerability in the coastal area of Bangladesh, thus emphasising factors such as gender, religious values, poverty, patriarchy, masculinity, unemployment, entitlement and access to resources, participation and involvement with decision-making process, power and infrastructural facilities. The theories drawn on in this study explore different levels of investigation ranging from the influence of socio-cultural and political norms of society to interaction between family members and finally an individual's characteristics. By drawing from the PAR theory by Blaikie et al. (1994) and Wisner et al. (2004); access theory by Blaikie et al. (1994) and Wisner et al. (2004); entitlement theory by Sen (1981), and theory of hegemonic masculinity by Connell (1995); and theories about social transformation and development (Beck-Gernsheim et al., 2003; Bradshaw, 2014), this study explored the potential for improving women's lives. These theories suggest that if women are mainstreamed in the development process by inclusion in the political process with the support of central and local GO along with NGOs, they will certainly develop agency and leadership capacity.

The political empowerment of women at both local and national level (the ratio of seats held by women in the national parliament increased from 10% in 1990 to 20% in 2011) and economic empowerment (employing over four million people in ready-made garments of which more than 80% workers are women) has ensured their growing presence in almost every sphere in Bangladesh (Basak et al., 2017; Chowdhury, 2015; Farhana et al., 2015). PAR lacks a sufficiently holistic theoretical analysis of complex power relations, which are a challenge for developing gender-responsive policies in this context (Moreno & Shaw, 2018; UN Women, 2016). Although there is criticism of PAR as a theoretical framework in explaining the progression and pathways of vulnerability of the women for socio-cultural reasons, there is no single blanket theoretical frame to explain the vulnerability of women in the context of a disaster. This study agrees with the argument of PAR that the disproportionate effects of disaster on women are socio-culturally constructed, which has been reflected in the most recent literature. PAR has been an effective tool for policy-makers and planners, GOs, NGOs and international donor agencies and development planners since its inception. PAR is being applied to explain the nexus among disaster, risk and vulnerability and to demonstrate how disasters disproportionately affect different groups of people and how disaster is rooted in socio-cultural conditions (Saha, 2015; Shreve, Davis & Fordham, 2017).

However, in the last decade, there have been changes in the nexus between socio-cultural conditions and vulnerability. For example, in recent times, changes in gender relations challenge gender relations further: *"Resilience and disaster can be a pathway to produce long term changes in gender relations and can empower women in the context of disasters"* (Moreno and Shaw, 2018: 205). Resilience has not been completely reflected in PAR. Recent research reports that: *"Women's vulnerability should not be assumed nor can the hazards they face be known in the current society... because of existing inequalities of power, women and girls may be the most impacted by exogenous shocks that bring both environmental and financial crises. Ironically, it is their lack of access to resources that may make them most practiced at responding to crisis through living this on a daily basis"* (Bradshaw & Fordham, 2014: 235). From this perspective, PAR explains that women experience different levels of vulnerability when gender intersects with characteristics such as class, ethnicity, sexuality, age and with life course but it is extremely linear; more rhetoric/speech-making and too generalised compared with action and practice in real life (Bradshaw and Fordham, 2014; UN Women, 2016). The social processes of change are important *"... the broad social process through which disaster[s]... are created adds to knowledge about the process of marginalization and social change movements towards inclusion and social justice"* (Enarson & Pease, 2016: 8).

Both access and entitlement theories are extremely important, which explain root causes of the disproportionate vulnerability of women who are not bound up only in socio-cultural assumptions. The restricted access and entitlements of women to assets and resources undermine the capacity to respond during a disaster and their capacity to be resilient (Bradshaw, 2014; Bradshaw & Fordham, 2014; Moreno & Shaw, 2018). Although the access and entitlement to resource approach has been used as an effective tool to explore the nexus between warfare, famine and vulnerability, it has been most applied in scholarly research and investigation in the field of disaster and natural catastrophes. Similar to PAR, the access and entitlement to resources-related approach has become popular among policy-makers, NGO activists, feminists, disaster researchers and scholars in recent times as a tool to explore the nexus among the lack of control over resources and disaster-related vulnerability (Saha, 2015). In this study, this theoretical framework has been used to explore whether access and entitlements to resources promote women in decision-making process at family and community level, the impact of migration and mobility and intersection with the sexual harassment of women. In other words, researcher was keen to use this theoretical frame to examine how lack of access and entitlements to resources in the coastal area creates disproportionate vulnerability to women.

In certain theories, resilience or the ability of people to respond and recover from disaster is inherent and latent (Birkmann et al., 2013; Moreno & Shaw, 2018). Writers emphasise that *"everyone has some capacity for resilience but not everyone can activate it due to limitation of access to and mobilization of, resources in responding to disasters"* (Moreno & Shaw, 2018: 206). The scarcity of food and scarcity of livelihood options as disturbances lead to a "positive functioning" (Moreno & Shaw, 2018: 206). This ensured women's access and entitlement to resources and access and entitlement to resources assisted women to have participation, developing agency and leadership in disaster management and cooperative action (Moreno & Shaw, 2018; Bhandhari, 2014). The ability of women to respond and their adaptive capacity shows that they are active agents in managing food storage, moving the household's living place higher during floods, caring for elders and sick members with indigenous knowledge, and caring for livestock and poultry (Saha, 2015). They can play an important role in *"caretaking, communicating risks, and organizing community activities"* (Moreno & Shaw, 2018: 207) involving women and building networks among women and building collective actions by women because they have easy access to women or inside the house (Gaillard et al. 2015; Saha, 2015; Shreve, Davis & Fordham, 2017).

However, despite the capacity of women to be active agents of change, a gap still exists in access and entitlement to resource-based theory. It does not explain how to reduce vulnerability of the women in the long term on its own. Moreover, gender theory is important. *"The manliness of men and boys is judged by their ability to measure up"* (Enarson & Pease, 2016: 7) and ideas about masculinity are slow to change but not impossible to change even in the isolated coastal society. The expectations of the impact of disaster on masculinity are changing. Connell's (1995) hegemonic masculinity, patriarchy and their intersection with other social divisions are not adequately addressed from a comparative/relative perspective, which considers the multiple axes of heterogeneity of social divisions, including religion, ethnicity, poverty, disability and power dynamics (Enarson & Pease, 2016). Considering the heterogeneity of social divisions for religion, poverty and power, researchers have applied theory relating hegemonic masculinity and patriarchy to exploring disaster in the context of long-existing patriarchy and masculinity in a traditional society but researchers require addressing certain changes in them. Connell (1995) explains that social and cultural acceptability of masculinity and patriarchy in non-disaster times puts males into a position of decision-making, setting down rules for family members, control over resources, possession of superior power and economic privileges. However, masculinity and patriarchy are endorsed, practised and tabooed in the context of social arrangements rather than a state of being (Bradshaw, Chant & Linneker, 2017; Enarson & Pease, 2016). Connell's concept of masculinity can be expanded to demonstrate more about "how the diversity of masculinities is marked by hierarchy and exclusion" (Enarson & Pease, 2016: 7). Their study shows how men respond to grief or stress after disasters in different approaches even within groups with similar class and ethnic backgrounds.

Disaster practitioners and disaster researchers report that Connell's hegemonic masculinity and patriarchy is too limited in recent times for analysing disaster since this is not the only element in play in the disaster domain. *"Connell's concept of hegemonic masculinity is not without critique ... charged with being unable to interrogate the subject deemphasizing power and domination, naturalizing the body and reifying power"* (Enarson & Pease, 2016: 7). However, this idea raised valid grounds that provide the conceptual tool to analyse disaster risk and vulnerability. Recently, it has been expanded as "women" and "men" "gender" and "sexuality", which are shown to be varied in disaster analysis. It can be used for interrogating men's practices in a disaster-related context (Enarson & Pease, 2016). However, the structure of Connell's hegemonic masculinity and dominance structure of the patriarchy became flexible/fragile in the coastal area of Bangladesh because of the "whole community" approach, "intersectional connection", continuous erosion of the role of institutional and socio-cultural support for both facing disaster-related risks and aspects of everyday life of women (Enarson & Pease, 2016). The involvement of women, engaging boys, men, young men, with community work, as well as social awareness, sensitisation of male members, social movements as an "intersectional and collective approach" ensured social inclusion and social justice for marginalised people such as women. The "intersectional and collective approach" effectively responded against the violence to women, including sexual violence, physical violence and domestic violence both during a disaster and normal time and then responded for facilitating gender equality (Enarson & Pease, 2016).

GO, NGOs, policy-makers, planners, donors and scholars focusing on disasters target women within households as deliverers of policy outcomes, policy shifts and practices for sustainable disaster risk management, disaster and popular culture, social justice, crisis management and reducing gendered vulnerability. Because masculinity and patriarchy are learned and marked by hierarchy that men and women internalise in their identities and women are brought up to care for others, women may be less defensive about change, and so it is easier to target them (Enarson & Pease, 2016; Bradshaw, Chant & Linneker, 2017).

Women headship became a rapidly accepted discourse and have greater visibility in the disaster-related context (Enarson & Pease, 2016). Female-headed households are a diverse group varying in respect of their composition (they might be headed by widows, or women who are divorced, separated, abandoned, and males are absent). They possibly emerge among very poor populations because of changes in the taboo on aspects of masculinity and patriarchy, which raise the prospects of women and helps them to exit from the poverty, social and economic disadvantages (Milazzo & Walle, 2015; Bradshaw, Chant & Linneker, 2017). After female-headed households experience fewer social taboos and fewer economic disadvantages, women can take the responsibility for their households on their shoulders (Bradshaw, Chant & Linneker, 2017). This leads to resilience and they can independently use their capacity. Policy directed at female-headed households draws on the concepts of empowerment and resilience but also ignores diversity among female-headed households and that it may be a temporary phase.

Marginalising women and keeping women away from development and intersection of race, class and gender, the *"matrix of domination, the power dynamics at play ... particularly when discussing individual's responses to disaster"* is examined in relation to the environment (Banford & Froude, 2015: 171). Although women may be disproportionately vulnerable via gender relations, their environmental knowledge on resource use and sustainable development helps community people's lives. This knowledge is shared via integrating/mainstreaming women in development programmes (Bradshaw, 2014; Banford & Froude, 2015). As an emerging tool in this study, feminist political ecology theory and ecofeminism explored the process of feminisation of work and gender division of labour in relation to nature as well as the environment. By mainstreaming women in development and using their indigenous knowledge and experiences of the environment to overcome constraints on their livelihoods, women can be especially resilient during a crisis. The involvement of women with environmental and development activism reveals how social, economic and political institutions and linkages operate (Banford & Froude, 2015). The political ecology of feminists explores the complexity of how women struggle for access to natural resources in different countries and over land, water and other resources where women have weak social position and security because of the interaction with existing social and political power dynamics as well as for religious and ideological viewpoints (Banford & Froude, 2015; Nibi, 2012). Nibi (2012) puts across the view *"that the struggle for resources (access and control) is not only negotiated by social relations of power but also through embodiments and emotions of people in different spaces and places"* (Nibi, 2012: 10).

The structural gender differences of knowledge, experiences and power because of access to development and environmental resources are shaping the adaptive capacity and resilience of women during a crisis. While disaster risk theory (Blaikie et. al., 1994; Tanner & Allouche, 2011; Wisner et. al., 2004) tends to address disaster and vulnerability and challenges in terms of socio-economic and cultural causes, the feminist political ecology theory and ecofeminism reconstructed the discourse that the vulnerability of women is associated with the process of development, modernisation and environment (Banford & Froude, 2015; Bradshaw, 2013), which has not been recognised so widely for a disaster risk theory. Currently, disaster researchers and practitioners know that gender is not only in play in disaster domains presumed to be gender-sensitive or in cultures presumed to be oppressive (Enarson & Pease, 2016: 8); rather, the access of women to development process and environmental resources minimises the risk and increases their adaptive capacity/resilience (Bradshaw, 2013).

In the context of disaster and women in the coastal area of Bangladesh, vulnerability is an extensively discussed component of scholarly research. Furthermore, in the light of disaster and vulnerability, social transformation and social changes are less discussed to a lesser extent. When this research started, it did as a usual attempt to unearth the vulnerability of coastal women; however, it subsequently revealed that there is social transformation and evidence of positive changes in the everyday life of the coastal women because of globalisation, economic growth, technological advancement of the country, in addition to the intervention of GO, NGOs and donor agents. For example, the image of the women as passive and their obligation to stay in the home and lack of agency have been changing because of democratisation. This might include ensuring the participation of women during an election, changing gender relations, women's empowerment, employment opportunities for women, activities by GO and NGOs, changes in the traditional decision-making process, advancement of satellite TV communication and mobile networks (Alston, 2015; Khan 2013; Panday, 2016; Rahman, 2014). This overarching transition of the society naturally impacts elements of women's life in the public and private sphere. It makes women increasingly more capable of transforming their surrounding environment and articulating their demands. As a group, women demand autonomy and they fight for equality to gain agency to influence and have control over lives. In addition to these factors for social transformation and social change, disasters as a factor contribute to changes in the social system (Moreno & Shaw, 2018).

1.9 Conceptual Framework

Because of the relevant available literature on the progression and pathways of the vulnerability of women, coping and adaptation mechanisms, changes in the role of women, empowerment of women, involvement and participation of women with decision- making processes and the role of GO and NGO for mainstreaming and developing the agency of women, this study has developed the following conceptual framework (Fig. 1.2). Moreover, this study has developed fewer models based on the results of the study, namely, progression and pathways of the vulnerability model (PPVM, see Chapter Three, Fig. 3.9), coping and adaptation model (CAM, see Chapter Four, Fig. 4.3), empowerment and change in the role model (DECR, see Chapter Five, Table 5.1) and discharging vulnerability model (DVM, see Chapter Six, Fig. 6.1) of the women for being able to take challenges because of developing agency.

1.10 Summary of the Chapter

The beginning of this chapter addresses issues, i.e., the traditional practices, lack of access/entitlements to resources to examine the discourse of disaster and vulnerability of women and their coping and adaptation mechanisms in the Bangladeshi coastal area. Then, this chapter discusses the role of NGOs as an alternative platform, and the policies related to disaster and climate change introduced by GO to explore women's leadership and mechanisms of GOs and NGOs interventions for including women in the decision-making process and any changes in the community as consequences of interventions by GOs and NGOs. There are helpful policies for initiatives such

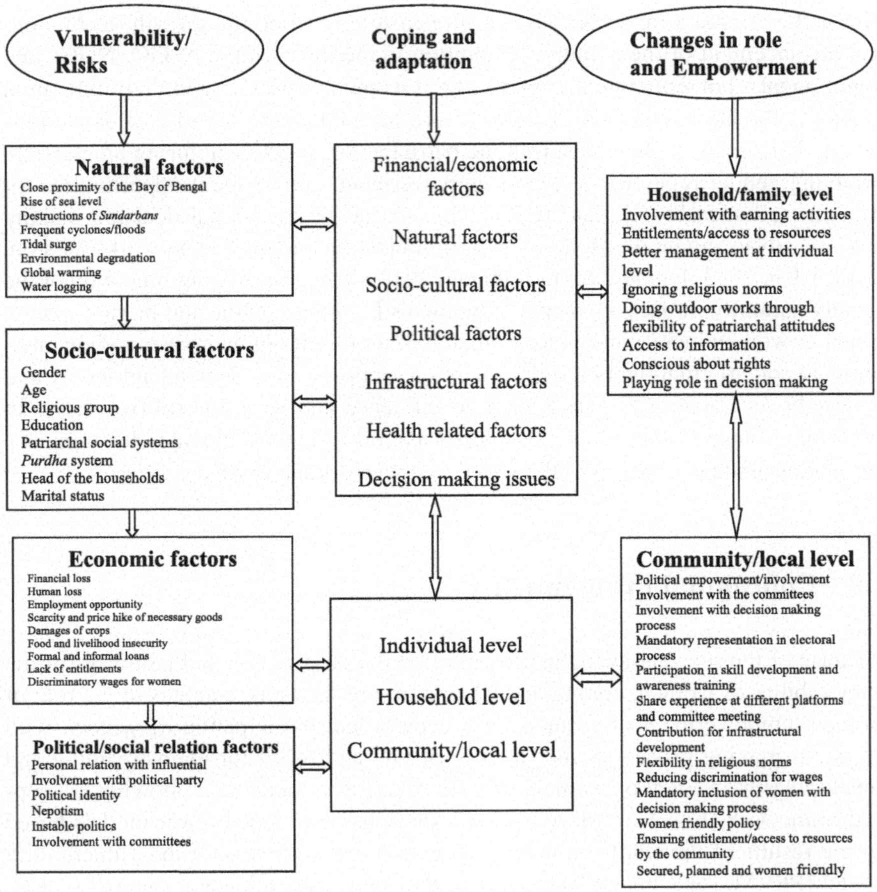

Fig. 1.2 Conceptual Framework

as advocacy programmes, capacity-building programmes, training programmes and planning frameworks. These initiatives are more only as policy in paper than practical implementation in the field, particularly by the disaster management institutions of the GO. However, the degrees of implementation and effectiveness of policies by both GO and NGOs are not equal. NGOs are more effective in implementing policies because of the nature of their activities, as well as their involvement and base of functions with people at the grassroots level.

In Bangladesh, what does the mainstreaming of women, the development of agency of women and the participation of women in decision-making look like in practice, one of the countries making significant progress on risk reduction, adaptation and capacity-building (Ayers et. al., 2014)? The risk reduction, adaptation, women's leadership for managing disasters and natural hazards in Bangladesh could have been better if, in addition to the effective implementation of policies, local contexts such as income diversification, addressing land and common property rights and access/entitlements to resources and management had been considered by policy-makers. Levina (2007) (as cited in Ayers et al., 2014) reported that access to financial markets and technologies improves the adaptation capacity of vulnerable people.

The long history of engagement in adaptation, the large number of organisations and agencies with knowledge, the GOs policies and the collective effort by GOs improved women's position in the coastal area by ensuring access to paid work, flexibility in the restriction of women going out, flexibility in following *purdha,* entitlements and access to resources. Flexibility in traditional practices and changes in lifestyle, livelihood, risk reduction, improvement of adaptation, capacity-building and resilience are visible in the coastal community. It is the outcome of implementing policies to disaster, collective effort of GO, NGOs and international NGOs.

International active participation of the newly formed Bangladesh government in World Conferences on Women in Mexico in 1975 followed by CEDAW (1979), Beijing Conference (1995), UN Decade of Women (1976–85), MDGs (2000–2015), SDG (2016–2021), focused on human rights for women and girls and drew international attention to climate change and disaster impacts. International attention helped the Bangladesh GO and NGOs to work together and speeded the progress in capacity-building, resilience and disaster management. For example, CDMP (a partnership project of GO with UNDP) has implemented policies and programmes for sustainable disaster management and resulted in changes in lifestyle and livelihood of women in the coastal community because of the transparency and accountability of GO to UNDP.

Although women are vulnerable in terms of taking food, household work, collecting water and wood for fuel, and going outside in the coastal area, the policies, programmes and projects undertaken by GO and NGOs have had little impact since 1971. Changes are taking place in society in terms of access to resources, in addition to the role and participation in decision-making and political participation. The knowledge gained by women attending different programmes, such as the training both from GOs and NGOs, not only effected change in the behaviour of women but also enhanced their understanding of the rights, roles and responsibilities of men. For example, women's capacity to raise their voices, as well as participate in earning activities and decision-making, at both home and outside, increased.

However, in the theoretical consideration, the PAR model explains disaster risk, in addition to the scale and magnitude of the hazard because of environmental factors embedded in the socio-cultural structures of the society, which produces vulnerability (Wisner et. al. 2004). The recognition of PAR as a model assists in collecting data in this study to explore the local context of disaster, women and vulnerability from macro, meso and micro aspects during a disaster on the distinctive features of existing social structure in the Bangladeshi coastal area. In this study, this model draws attention to how the social, religious, economic and political characteristics create pressure in determining the everyday life of women. Therefore, vulnerability results from different social factors and lack of access to resources that determine increasing disproportionate socio-economic suffering and loss because of climate change.

To obtain relief from the pressure, the nature and pattern of coping mechanisms and adaptation strategies need to be considered, which will in turn contribute to resilience; otherwise, it will counter the effects of vulnerability (Blaikie et. al. 1994; Wisner et. al., 2004). The access and entitlement approach expands the PAR model and demonstrates that lacking access/entitlements and control over available resources in a given society is associated with causing unsafe conditions and disproportionate vulnerability for women. This study receives guidance from the access and entitlement approach to explore how control over available resources is ensured and how people manage vulnerable conditions, levels of risk, coping and adaptation mechanisms to disaster at individual, household and community levels.

This study receives guidance from the theory of hegemonic masculinity, which expands on the gender aspects of the framework. This theory confirms what makes men dominant and by what mechanisms this is maintained and challenged. Recognising the theory of hegemonic masculinity, this study attempts to explore the power dynamics of men over women that are supported by structural inequalities. This allows the researcher to discover the challenges for women such as the practice of patriarchy and domination of masculine attitudes to understand the gender relation in Bangladeshi society, which has extended in this study by focusing on feminist political ecology. Feminist political ecology theory has been considered in this study for more explorative study of gender and gender relations to nature/natural resources as a force for understanding the distribution of power dynamics in a given society.

Social transformation theory addresses that over the last 20 years, women as a group became capable of transforming their status, and began to demand autonomy and equality. The emergence of strong NGOs and women's movement after the 1980s promoted activism, the political voice of women and foundation of women's human rights in Bangladesh in terms of access to education, technology, power distribution and involvement with income-generating activities and mobility. This study explores social transformation and mainstreaming of women with development programmes, and the participation of women in the decision-making process at both family and community level. It also looks at the intervention of GO and NGOs to develop the agency of coastal women, direct participation with the political process, working outside the home, earning, doing business in a public place (see details in Chapter 4, Chapter 5 & Chapter 6), the situation of domestic and sexual violence, and mobilising the community for women's rights in the coastal area.

Chapter 2
Methodology and Sources of Data

2.1 Introduction

Methodology is one of the approaches in which the research process is made transparent; it can be used as a strategy or a plan of action. Methodology provides grounding logic and assumptions before the design is finalised (Silverman as cited in Phellas, 2006). It helps understand why this study has been undertaken and how the research problem has been defined, what type of data has been collected, what particular method has been adopted and why a particular technique of analysing data has been used and a host of similar questions. This chapter discusses the issues related to the methodology of this study such as research design, methods of data collection, access to the research site, sample design and data analysis to measure the research questions and research-related phenomena. This chapter discusses the process of execution of the design of the study, justification of mixed methods, problems of the field research, limitations of the study and how limitations have been overcome to obtain reliable and valid data.

2.2 Rationale of Mixed Methods Triangulation

For an improved understanding of disaster, gender relations, gendered vulnerability and changes in daily practices in the coastal area in Bangladesh, a mixed method methodology research design was selected. Mixed methods are able to develop knowledge on particular phenomena to get insight and in-depth information even for certain gender-sensitive issues. The quantitative data of this study was collected from coastal women using a survey method with a structured questionnaire with a few open-ended questions. It elicits data on the impact of events and ownership of property, which requires a systematic approach.

Qualitative data gains improved descriptive understanding about the experience of vulnerability. Therefore, focus group discussion (FGDs) and in-depth individual

M. M. Khan, *Disaster and Gender in Coastal Bangladesh*,
https://doi.org/10.1007/978-981-19-3284-7_2

interviews were conducted among different groups of coastal people with certain consisting of men only, some of women only, and some a combination of man and woman. Single sex groups allowed those details that are sensitive in the context of societal norms to be collected that respondents do not feel comfortable to answer in the face-to-face survey questionnaire.

Although there are multiple arguments and debate about why mixed method research design is considered as a separate research design in social sciences (Cresswell et al., 2003), this method was extremely useful and effective in determining the discourses of disaster, women and vulnerability in the coastal area in Bangladesh. In this study, the mixed method has been applied as this is pragmatic and fits with the aims of the study to help policy-makers understand the coastal study community (Tashakkori & Teddlie, 2003). The forms of vulnerabilities, nature of gender discrimination, violence, local politics, forms of deprivation, local practices and masculine attitudes could be understood better via qualitative lens.

The application of mixed method triangulation helped generate applicable knowledge from this research better that that for further planning for the people of the disaster-prone coastal area. FGDs and in-depth individual interviews with different stakeholders and members of different committees as a qualitative tool were an extremely effective approach to gain insights and in-depth information on gender-sensitive issues and to have views of men regarding gender relations, vulnerability of women and changes in practices.

Hence, motivated by the theory of PAR, access theory and entitlement approach, theory of hegemonic masculinity, feminist political ecology approach and theory of social transformation, this research has been conducted around the following research questions such as how do disaster, women and vulnerability function in disastrous times in the Bangladeshi coastal area? What are local practices that are causing disproportionate vulnerability for women? How do unequal access and entitlements to resources impact women in vulnerable situations? How does involvement and participation of women reduce vulnerability, develop agency and empower them? How are GO and NGOs interventions contributing for reducing vulnerability and mainstreaming women with a decision-making process? Are there any changes in local practices, cultural context, lifestyle, livelihood options and participation with decision-making process for women in facing disaster?

These questions can be answered by understanding the flow of information and getting information from both men and women of the coastal community. Researchers were interested in identifying the dynamics and mechanisms of society that intensify the vulnerability of women and in understanding the mechanisms and behaviour in disastrous times and changes of practices for women. Methodological triangulation/mixed method allowed researchers to cross-justify the reliability and validity of inquiry. Using the mixed method, reliability and validity are improved and researcher has enlarged the understanding of the range of factors, which required to be in an account to understand and explain the social relations and behaviour observed (Nasreen, 1995).

2.3 Socio-Demographic Introduction of Research Settings

This study has been conducted in Sutarkhali and Kali Nagor villages of Dacope *upazila* in Khulna district, Bangladesh (Map 2.1). Dacope upazila in Khulna district is almost 25 km away from Khulna city. It has nine unions, namely, Bajua, Banishanta, Laudubi, Chalna, Dacope, Kailasgonj, Kamarkhola, Sutarkhali and Tiladanga. The total area of Dacope is 992 km^2. The total population is 157,489: male 53% and female 47%. The mixture of religions is typical of the Bangladeshi coastal region: Muslim 42%, Hindu 57% and Buddhist 1%. Similar to other villages of Bangladesh, the study villages have their own places of worship. There are 72 mosques, 137 temples,

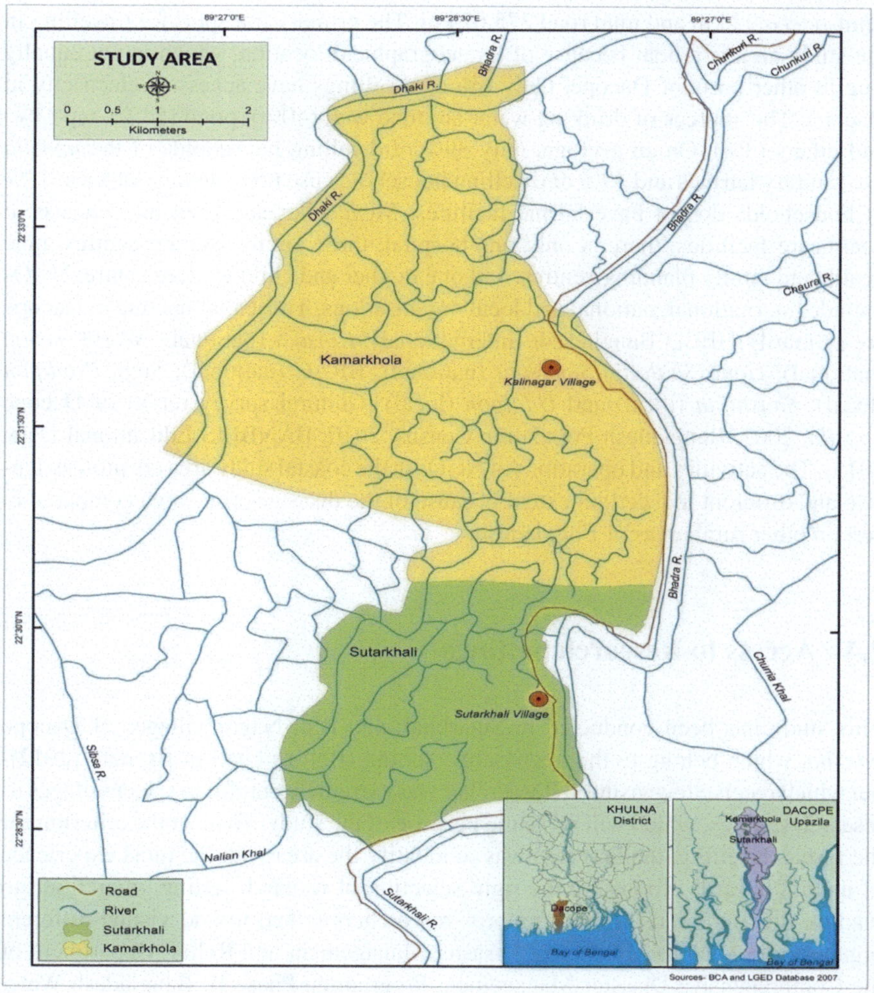

Map 2.1 Location of Sutharkhali and Kali Nagor village (Research Settings)

11 churches and two sacred places in the whole *upazila*. There are six colleges, 31 secondary schools and 56 primary schools along with 43 NGO schools and five *madrasas*; these educational institutions are scattered throughout the *upazilas*. In every village, there is at least one primary school. The rate of male literacy in Dacope is 54%, whereas female is 39% compared to the female rural literacy rate of 44% at a national level. The dropout rate in the study area is 20% (21% at national level), 19% for girls and 20% for boys (Cultural survey report of Dacope *upazila*, 2007; Bangladesh Population Census, 2010).

Similar to other rural areas of Bangladesh, the primary income sources are agriculture (66%), commerce (13%), service (4%) and non-agricultural labour (5%). The ownership of agricultural land in Dacope is distributed as landowner (55%) and landless (45%). The roads and communications are *pacca* (brick built) only 18 km, semi-*pacca* 92 km and mud road 275.67 km. The primary transport for travelling in the study area is a boat because of its geographical location, which is not equally true in other parts of Dacope. Only 6% of dwellings have access to electricity in Dacope. The sources of drinking water are tube well (40%), pond (54%), tap (3%) and others (3%). On an average, only 39% of dwelling households of the *upazila* use sanitary latrines and 49% of dwelling households use non-sanitary latrines; 13% of households do not have latrine facilities. Most villages do not have access to healthcare facilities, there is only one hospital, three family welfare centres, nine health and family planning centres, and one mother and child welfare centre. NGOs include international, national and local organisations. Those that operate in Dacope are primarily HEED Bangladesh (international), *Prosika* (national), World Vision (national), *Gono Shahajjo Sangstha* (national), BRAC (national), Step, *Prodipon* (local), *Shushilan* (local) and *Uddipon* (local) (Cultural survey report of Dacope *upazila*, 2007; Bangladesh Population Census, 2010; BANBEIS Educational Data, 2015). The activities and operations of NGOs in the coastal study area are more extensive and different to a certain extent because of the disaster-prone area compared to certain other rural areas of Bangladesh.

2.4 Access to Research Settings

This study has been conducted in Sutarkhali and Kali Nagor villages of Dacope *upazila*, which belong to the open/public setting (Bell as cited in Bryman [2012]) but which receive few visitors. Researcher was extremely careful to select villages as research settings, which will serve the purpose of the study. To meet the criterion for the research setting, the first step was to identify the area with the most experience of natural hazards. To ensure the right selection of research setting, in addition to reading literature and different reports, researchers talked to and visited different organisations such as Ministry of Disaster Management and Relief, Department of the Comprehensive Disaster Management Programme Phase II, Bangladesh Water Development Board and different NGOs operating different programmes in disaster-prone areas. Researcher chose Sutarkhali and Kali Nagor village as a study setting bearing in mind that villages and villagers have experienced natural hazards such as cyclone, tidal surge, bank erosion and flood.

The study area is located very close to the Bay of Bengal and is always at the risk of natural hazards, including cyclones, tidal surge, flood and river bank erosion. Both villages were severely affected by *Aila* in 2009, the most recent severe cyclone in Bangladesh; in the last few years, Sutarkhali and Kali Nagor were affected by the devastating cyclones *Sidr* and *Nargis*. Sutarkhali and Kali Nagor were deliberately selected for this study considering certain predetermined reasons such as: (1) these two villages are situated very close/adjacent to coastal stop banks, (2) the two villages have very easy and direct access to water through the *Shibsha* river, (3) these two villages are just behind the embankment, and (4) these villages have experienced the most devastating and extreme damage from *Aila* compared to other villages and (5) villagers continually worry that a disaster will break up the poorly constructed local dam.

One of the important and most important steps in research is to gain either overt or covert access into the research setting in which the researcher is interested (Bryman, 2012). Access of the researcher to villages was overt for this study as the researcher followed a top-down access route to the research setting to ensure easy access rather than following a bottom-up access route. As a researcher access to the research setting was not so easy and straight forward because the research site was ~ 600 km away from the city the researcher lived in.

The research site, in addition to the people of the community, were totally unknown to the researcher who knew it would be difficult if the researcher went over there straight away to conduct fieldwork and interviews. To make access easier, the researcher met certain GO officials in Dhaka and explained the aims and objectives to them, thus showing researcher's business card and a letter from the Jagannath University. Fortunately, the researcher managed to meet the head of the CDMP II programme. He introduced the researcher to another official of CDMP II who had visited Sutarkhali and Kamarkhola union many times for projects and activities related to the disaster. The Head of CDMP II made a phone call to Dacope *upazila* PIO and requested help for the researcher to conduct research. The other CDMP II official gave his phone numbers to the Chairman and the elected representative of Sutarkhali and Kamarkhola Union; therefore, the researcher could talk to them with the recommendation of the CDMP II programme's head.

For the first time, when the researcher went to Dacope met the UNO and PIO, introduced himself in person and sought help and assistance. Both UNO and PIO made a plan for researcher to go Sutarkhali and Kali Nagor village. They made a phone call to the union chairman, and a GO official accompanied the resesearcher the following day on a field visit. The GO staff who accompanied researcher introduced researcher to the chairman and others in the Sutarkhali and Kamarkhola union. On the first visit in the field, the researcher met the chairmen of both Sutarkhali and Kamarkhola union and certain other local leaders. The chairman made several phone calls on behalf of the researcher and introduced over the phone to local leaders and influential people of Sutarkhali and Kali Nagor village. This is how researchers built a network and relationships with the people of different categories of the local area, including elected representatives, GO officials, local political leaders, NGO activists, religious leaders and school teachers. Furthermore, all steps have been followed by the researcher only to get access to the villagers and be oriented with the villagers such that villagers get confident to share information and data with researcher and research assistants.

Before starting pilot and data collection, researcher visited villages and organised multiple orientation programmes with community people to inform them about the purpose of this study and build a rapport with them. Building a rapport and getting access in a traditional conservative society was extremely difficult as an outsider but people of all walks of the community did not differently think, although researcher gained access to the site via rich and influential people. Researcher was able to build a rapport with community people because he spent six months there doing fieldwork. It was not easy to leave the research setting at the end of research as the researcher had built up a friendly relationship with certain people in the area.

2.5 Population and Sample of the Study

For this study, the "universe of units" (Bryman, 2012) for quantitative data was only women who have experienced and been affected by different disasters. Data were collected from groups with varying socio-economic, political, cultural and professional status. In this study, the method of drawing a sample was a random choice from selected family members. Initially, researcher had a plan to interview respondents by applying the systematic random sampling method based on the researcher's knowledge of the structure of villages in Bangladesh. However, the structure of villages in Sutarkhali and Kamarkhola union was different to usual structure of villages in other parts of the country. Both households and families of Sutarkhali and Kali Nagor were scattered because of the nature of local geography. The locations of households were on available high land, available communication facilities, secured protection from tidal surge and flood. It was extremely difficult to collect data from households by applying systematic sampling techniques.

At the start of data collection, researchers decided to stay in villages and proposed this to villagers. However, there was no guest house in villages where someone could stay for such a long time period during fieldwork, although certain villagers insisted to stay with them. Finally, it was impossible because most households have a maximum of one or two rooms and that is not even sufficient for the members of the household; therefore, they were not in a position to accommodate researcher and research assistants. Moreover, researcher reported there was no electricity connection in those villages to work at night after each day's data collection. Note that only 1.77% and 6.40% of households (BBS, 2010) are connected via electricity in Sutarkhali and Kamarkhola union, respectively.

Considering the circumstances, researcher determined to stay in the *upazila* headquarters, which was 2–3 h journey by *trawler*. The *trawler* is the main and extremely common medium of transport in Sutarkhali and Kali Nagor. The time of journey to reach the study setting depending on whether the boat was going with or against the river current. The research team travelled every day to the research setting and came back to *upazila* headquarters after every day's data collection; moreover, the *trawler* was only able to go where there are rivers or canal ways. Because of the unusual geographical structure of villages, villagers were used to the living scattered only in the high land area. In this study, each community of high land has been considered as a cluster. For extreme problems of communication in the research area, the researcher decided to use random sampling for quantitative data and respondents were randomly picked from each cluster.

In this study, each household was considered as a sample unit; from each household, only one woman as a respondent was surveyed. Before data collection, it was decided to interview ~ 200 respondents to be adjusted depending on the time frame, funding, number of households and age group. Once the researcher decided on the sampling procedure, he was able to increase the total number of respondents to 263 respondents. The researcher decided to interview respondents aged > 15 years because of the nature and aims of the study. The total number of respondents, FGDs and in-depth individual interviews was appropriate for this study because the population of the study area is homogenous in terms of their occupation, religion, education, lifestyle and disaster experience.

For qualitative data, a total of nine FGDs and nine in-depth individual interviews were conducted among disaster-affected men, women, elderly women, female household heads, NGO activists, political leaders, community leaders, elected representatives and with different committees, which manage the local response to disasters. The venue and time for FGDs were selectedto be convenient to participants. In each FGD, the maximum number of participants was 15 and the minimum was 5. Certain FGDs were only with women participants because of certain sensitive issues to be discussed with women. There were no women participants while FGDs were conducted with NGO activists because of inconvenient time, socio-cultural practices and transportation. FGDs with NGO activists have been conducted at ~ 9 PM, which is extremely late and not considered acceptable for a woman to be out in a village. It was not only because of socio-cultural practice, transportation was a problem. NGO activists could not manage FGDs during the day time working hours. The participants of FGDs were confirmed with the knowledge of both participants and their guardians or heads of households. Participants had complete freedom regarding participation and they intentionally attended because they knew the time and venue and they determined whether to participate in the discussion.

The site for FGDs with UDMC was the union *parishad* office. However, FGDs with vulnerable women and CPP members in Sutarkhali were in the villagers' social club, which is situated in a place termed by the community people as *mor* or a common place/room for community people. Nevertheless, in Kali Nagor, an FGD with vulnerable women was arranged in a place close to the local *bazar*; however, the FGD with female-headed households in Sutarkhali was conducted in the house of the president of PIC because there was insufficient space in his house and everyone knew the house. However, the president himself was not present during the FGD. Only the

FGD with NGOs was arranged in the *upazila* bungalow, the place researcher have stayed during fieldwork, which was far away from research setting.

The place and time for in-depth individual interviews was fixed, thus giving priority to opinion and convenience of interviewees at their home, a tea stall or work-place. The respondents for in-depth interview from the most vulnerable community people (avoided influential and known people for in-depth interview) were appointed over mobile phone communication before going to the research field, although these arrangements were interrupted because of the local election campaign and the busy season of farming.

Participants completed a consent form before starting the FGD. Before each FGD, a brief introduction about the study was read. The smallest FGD comprised 5 members and the largest was 15 members. A total of 46 women and 49 men participated. The age range of participants was from 18 to 80 years. Note that the researcher had no freedom to consider the age of the participants for conducting FGDs with UDMC, CPP, PIC, NGOs or female-headed households. However, for the FGDs with vulnerable women and community people, the researcher prioritised young, adult, knowledgeable, well-informed, vocal and active participants. When conducting FGDs, researcher attempted to avoid any biases in who attended, such as political and gender; however, it was outside control of the researcher because almost all FGD participants were organised with the help of local people and locally elected representatives. All participants of FGDs and in-depth interviews willingly participated; furthermore, the number of withdrawals and those leaving in the middle was zero because FGDs were arranged with their consent.

The participants were encouraged to discuss the relevant issues in a naturalistic and conversational manner (Saha, 2015). Each FGD lasted on an average 1.5 h and was attended by a note taker, and the researcher acted as moderator and followed the conversation flow as per the pre prepared written checklist. Usually, discussions in FGDs are tape-recorded and then transcribed (Frith, 2000); in this study, the discussion was tape-recorded on two different digital MP3 players (to avoid technological difficulties and in case of failure of one) and transcribed using Microsoft Windows Media Player and then translated from Bangla to English.

2.6 Data Collection Techniques

The field research was performed for six months, including the pre-test and orientation of the research and research teams with the community, consultation programme with the community people, and network and rapport building with the local community. A pilot survey was performed to verify the sequence and significance of questions and aptness of language with non-sample similar respondents in two other villages in the same coastal area. After the pilot test, a few questions had to be changed, withdrawn and included in response to demands of actual requirement of field respondents in the final survey interview questionnaire. Then, the final questionnaire was translated into Bangla, the native language of the research area.

The research team attempted not to be disruptive of the respondents' everyday lives and talked with the respondents wherever they were working. Research assistants were asked to ensure respondents knew there was no pressure to answer questions and respondents could withdraw part way via the interview if they wished. Research assistants were seated on the ground beside them, and research assistants were trained up to address the respondents in kinship terms such as *bhabi, boudi, chachi, mashi, apa, didi, dadi* and *nani* to establish relationships with them. Before interviewing a respondent, the interviewer presented the questionnaire to be read by those who could read; moreover, it was read by the interviewers to respondents who could not read. They discussed consent with the respondents who signed a consent form prior to interview. During this study, it was farming time in the locality. However, despite the repondents' busy schedule, they showed considerable patience in responding to questions. All face-to-face interviews were conducted at the house of participants where they could feel comfortable.

In addition to face-to-face interviews, the researcher collected meeting minutes and meeting regulations of UDMC from the union *parishad* office (office of the union chairman). The meeting minutes and meeting regulations provided information about the number of female members in different committees, the rate of participation of women in the decision-making process, and the involvement of women with different activities and programmes in the local area. The meeting minutes and regulations were a reflection of the women's attendance and helped the researcher cross check the information researcher obtained from the field research.

For qualitative data, FGDs ("such as group interviews" or "like-minded people discussions" to gather diverse opinions, experiences, beliefs, feelings and attitudes [Frith, 2000]) and in-depth individual interviews have been conducted for shared ideas and experiences from people of different stages in the coastal area of southern Bangladesh. In this study, the people of coastal area who experienced disaster such as sidr and *Aila* in their life were considered as a "like group" and "like-minded people". Moreover, FGDs were conducted in the light of the concept of "a circle of chairs" (Chadwick, 2002) where in the circle, all participants were equal and there was no head of the table to set the power figure apart and everyone could see everyone. There was no behavioural message to show someone was more important FGD usually began with self-introduction by the participants. Subsequently, there was an overview and explanation of the study purpose by the researcher, followed by a question to describe the experiences they had during the last cyclone, including problems, coping mechanisms and gender-based problems. Then, the group discussion proceeded around the circle and researcher attempted to maintain a time limit and discuss all issues mentioned in the checklist. All tape recordings were transcribed in Bangla by the researcher and research assistants during the field research and then translated in English by the researcher.

Purposeful sampling strategies (Cresswell, 2003, 2013 as cited in Saha, 2015) were used for selecting FGD participants and in-depth interviews that facilitated understanding the gender, gender relation and gendered vulnerability in coastal Bangladesh. As part of participatory approach for qualitative data, we carried out different FGDs with vulnerable women: one with female-headed households, one with vulnerable community people, one each with the members of UDMC from two different unions, one with PIC members, one with CPP committee members and one with staff of NGOs (Table 2.1).

For qualitative data, in-depth individual interviews were conducted for rich information on personal experiences, ideology, social structure and institutions with a semi-structured checklist. In the checklist, the researcher organised the questions into broad topics such as experience in daily life, entitlements, empowerment, migration and leadership; moreover, a number of sub-topics were prompted within each topic. The respondents were free to flow from one to another and use a different order and ensure all areas of the interview were covered, although it was difficult because a semi-structured interview is driven both by respondents and interviewers. Each in-depth interview was recorded with the prior consent of the interviewee. After each interview, the researcher assessed the success of the way each area had been covered or not and, if required, the researcher added and subtracted or reordered the points on the interview schedule before the subsequent interviews. In-depth interviews were between 45 min and 2.30 h. Of these, only two were < 1 h. Shorter interviews were constrained by respondents' busy schedules, particularly those that were taken at work places. Researcher worked hard to conduct interviews in a private place where no one would interfere and respondents felt comfortable and secure to share sensitive issues and obtain thoughtful answers.

2.7 Statistical Analysis

Data analysis stage incorporates multiple elements and raw data has to be managed to establish whether there are any obvious flaws such as failure to accurately answer (Bryman, 2012). Analysing data is an elaborate process of reducing raw data in concepts by transcribing, thematic analysis, coding data, writing memos, diagramming and the process of application of statistical techniques by a computer to examine what they can say about research questions and hypothesis (Bryman, 2012; Neuman, 1997; Punch, 2005). In the field, the research assistants were instructed to write a short note if there was a long answer and opinion on anything interesting from the respondents to avoid any discomfort and boredom for the respondents.

Each night after coming back from the field, the researcher sat with research assistants to understand their experiences in the field and for data cleaning and removing errors (Rahm & Do, 2000). The closed questions were pre-coded, and the code number was fixed after pilot study. The answers to open-ended questions were coded at the end of field research. All open-ended answers have been separately written against each question, and consequently the answers were sorted out and matched with the different categories of answers. After different categories of answers have been sorted out for open-ended questions, the researcher coded the answers as the basis of interpretation (Bryman, 2012). To analyse and interpret coded quantitative data, IBM Statistical Package for Social Sciences 20.0 (SPSS) has used in this study. The researcher presented the results of the study with the literature and have shown

Table 2.1 Socio-demographic descriptions of FGD participants

Location for FGD	FGD group	Age range	Total participants	Participants by gender		Participants by schooling year	
				Men	Women	Schooling year	Participants
Model Village[1] (Sutarkhali)	Female-headed households	30–80	13	-	13	Sign only Some schooling	5 8
Social club (Sutarkhali)	Vulnerable women	22–42	10	-	10	Sign only 0–5 6–10	3 3 4
Social club (Sutarkhali)	CPP members	21–62	15	12	3	0–5 6–15	3 12
Gazi para (Sutarkhali)	PIC members	32–55	8	7	1	0–10 11–15	4 4
Sutarkhali UP	UDMC	28–64	13	9	4	6–10 11–15 16 +	7 4 2
Khamarkhola UP (Kamarkhola)	UDMC members	30–60	11	8	3	0–10 11–15	6 5
Telirkona Bazar (Kamakhola)	Vulnerable villagers	18–75	9	8	1	Sign only 0–5 6–10 11–15	2 2 2 3
Joy Nagar Bazar (Khamarkhola)	Vulnerable women	19–55	11	-	11	Sign only 0–5 6–15	2 5 5
DacopeUpazila Bungalow	NGO activists	34–42	5	5	-	11–16 +	5

[1] A village has been established by the government on abandoned land after *Aila*, with a particular structure of house and a tubewell, sanitation facilities, water reservoir and community pond for people who lost their houses in Aila. The local people called it a model village.

dimensions of relations between variables via descriptive quantitative analysis such as graphs and tables. This study has rounded up percentage (%) in complete numbers rather than keeping fractions in the text.

To analyse qualitative data of this study, the researcher narrowed down the core themes of the study based on research questions using axial coding, the process of relating categories to the sub-categories to form more precise, and complete explanation about phenomena from different level of properties and dimensions (Strauss & Corbin, 1998). Strauss & Corbin (1990, 1998) propose that coding and categorising data are related to (1) cause and effect conditions, (2) the phenomenon of study, (3) the context conditions, (4) the intervening conditions, (5) actions/interactions strategies and (6) consequences.

In this study, axial coding is used for qualitative data analysis because it helped to condense voluminous qualitative data (Corbin & Strauss, 2008). Moreover, certain qualitative data is presented in a descriptive mode as quotations in this study. Based on the written transcripts from audio recordings, seven broad categories based on research questions and sub-categories were generated (Table 2.2). However, by axial coding techniques, this study focuses on socio-cultural, economic, geographical and institutional vulnerability, vulnerability of female-headed households, access and entitlement to resources and vulnerability, women's involvement and participation with decision-making processes and committees and vulnerability, changes in local practices, and adaptive capacity and coping mechanisms from different dimensions considering context conditions, intervening conditions, structural conditions or causal conditions (Strauss & Corbin, 1990, 1998).

The researcher carefully repeatedly read the transcripts to identify connections between the participants' discussion and research questions. In the FGD transcripts, there were blocks of information related to the checklist for FGD. The in-depth individual interviews were divided in different blocks. These data blocks were analysed to discover themes, which is termed as sub-category (Corbin & Strauss, 2008; Strauss & Corbin, 1990, 1998), such as migration and vulnerability of women, disaster and violence against women, disaster and health vulnerability of women, patriarchal norms and vulnerability of women, entitlements and access to resources and vulnerability of women, food distribution and vulnerability of women, disaster and vulnerability of female-headed households, adaptive capacity, coping mechanisms, resilience and women's participation. As per this procedure for qualitative data, the researcher explored themes and patterns emerging from the data and incorporated these with quantitative interpretation, which strongly bridged between quantitative and qualitative data and enriched the explanations of research results.

2.8 Validity and Reliability of the Data

Both reliability and validity are releated to how concrete measures and indicators are developed to "maximise the reliability and validity" of the study (Neuman, 1997, 2006). Fundamentally, reliability is related to issues of consistency of measures of

Table 2.2 FGD and in-depth individual interview analysis following axial coding procedure

Core themes/category	Sub-category
Vulnerability	**Social-cultural:** Displacement and migration, homelessness, insecurity and violence, sexual abuse and violence, corruption, uncertainty and anxiety
	Economic: Scarcity of food, damages of livestock, poultry, shrimp *gher*, agricultural land, assets, poor income, shattering employment opportunities, dependency for aid/relief
	Institutional: Limited safe shelter centre, inaccurate warning system and lack of advance warning systems, inadequate arrangement for water and cooking, lack of shelter centre for livestock and poultry, inadequate sanitation and healthcare facilities
	Geographical: Coastal proximity, destruction of Sundarbans, tidal surge, mighty river, riverbank erosion, weak embankment and dam, suitable for shrimp farming
Female-headed households and vulnerability	Social taboos, ignorance, undervaluing, exclusion, insecurity, wage discrimination, relief/aid discrimination
Women access/entitlements to resources, involvement with income generating	Access/entitlement to resources and properties, involvement with income generating, access to loan/micro credit/instalment payment, dependency of women, decision-making, bargaining capacity, power to spend money
Participation of women and better disaster management	Representation of women in committees, barriers for inclusion, role and evaluation of opinion in the meeting
Bottom-up approach	Mainstreaming women, ensuring local participation, accountability and sustainable development, minimising gaps
Changes in local practices	Involvement with income-generating activities, relaxed in practising *purdha*, freedom of women in mobility, role in decision-making process, empowerment and voice raising
Adaptive capacity and coping mechanisms	Different response for different backgrounds, unemployment, health care, *purdha* system, foods scarcity, sexual violence, insecurity, difference between adaptive capacity of men and women

concept that involves stability, internal reliability and observer consistency. It informs us about an indicator's dependability and consistency (Bryman, 2012). However, validity is related to the truthfulness and trustworthiness of the conclusion and results that originate and are drawn from this research. In other words, validity is the condition of having really measured what one wanted to measure (Bryman, 2012). Both reliability and validity are concerned with the adequacy of measures, which are most obviously a concern in social research (Bryman, 2012). However, validity cannot be achieved without the research first being reliable. Moreover, reliability and validity are important issues in any research project (Cresswell, 2007).

This study has been conducted with the highest effort and careful attention to ensure and maintain the consistency in the data and explanatory arguments. To maximise the reliability and validity of data in this study, the interview time was fixed based on the availability of respondents and time they gave for interview. Moreover, the respondents had freedom to withdraw themselves from the interview at any time. Furthermore, research assistants were asked to stop work if they were concerned about something or too tired. There was no target given to them to complete the questionnaire within a certain time. All members of research team used to meet, most possibly at a local *bazar* or tea stall, for refreshments after working for a certain time and to share any problems or experiences they had. As a researcher, surprise visits were conducted to research assistants without any prior notice or phone call to ensure the quality of work.

Although there are multiple threats to validity and not always controllable or even perceptible (Blalock, 1985), to maximise the accuracy and credibility of data, the researcher has applied methodological triangulation to confirm the information. The researcher used to review every day's data and verified the accuracy of survey result with FGD. The researcher undertook informal conversation in the field with community people about trends in the previous day's results that we had discussed in a group at night. The researcher confirmed and cross-verified certain results multiple times with the community residents whenever the researcher doubted the reliability of specific data that he heard from research assistants because the complex nature of the social phenomena makes validity harder to gain (Blalock, 1985). To ensure higher reliability from FGDs, tape recording has been used with the consent of participants. The researcher was actively involved with this study in the research site and this was ongoing for six months. The experience of working in the research site for six months in person helped to identify issues and concerns on factual occurrences.

2.9 Limitations and Validity Threats

There are many limitations in this study. Prior to starting fieldwork, the researcher had only certain preconceived theoretical ideas relevant to the research objectives. The research assistants had no in-depth knowledge about this research. Moreover, research assistants were trained based on the survey questionnaire, FGD and in-depth individual interview checklist. Consequently, they sometimes faced problems while interviewing respondents, particularly for the first few interviews.

Although research assistants were asked to make a mobile phone call to the researcher with questions, sometimes because of network failure and weak signal they could not reach the researcher at the right time. Moreover, there was discomfort of respondents with certain parts of questions such as sexual content, male domination and patriarchal system contents. Sometimes, survey respondents avoided answering some questions because of the presence of the male member of the household at home. Moreover, particularly during mixed group FGDs, where men and women participants avoided responding to certain sensitive issues. Women discussants did not feel it was in their interests to say anything negative, e.g., regarding the distribution of relief and role of committees and sexual issues in the presence of male discussants and elected male representatives.

The researcher was very much targeted on research questions and objectives because the fieldwork cost was financed by the researcher. Furthermore, financial and time limitations forced the researcher to complete the fieldwork within six months. Because this study was very context-specific, the results may not be applicable to other countries and even to other regions of Bangladesh because of different biophysical, socio-economic and cultural contexts. As a researcher, it was important that a rapport and network was developed within the community. The researcher always aspired to improve rapport and networking; therefore, the research would be better and would like to stay longer in the field. However, the researcher worried that sometimes people might over identify with researcher's ideas at the same time.

An information sheet was distributed, and the contents of this study were discussed and read out among respondents and discussants prior to every face-to-face interview, FGD and in-depth individual interview. To avoid irrelevant issues, respondents and discussants were requested to only answer the relevant question and give their opinion and share their experiences. However, while interviewing respondents, other women members often answered the question on their behalf. Sometimes, a male member of the family was attempting to answer different questions, and it was extremely common to have interruptions.

In FGDs, many women participants felt uncomfortable in sharing experiences involving sexual matters, exploitation and discrimination matters between man and woman. Although the researcher conducted a few FGDs only with women, the researcher felt that a few more FGDs with women could give better information because women feel free and comfortable sharing any issues in the presence of women. Although FGD participants were requested not to talk on irrelevant issues, there was always a chance they talked on political issues. Direct discussions on political issues become unpleasant for other participants even for researcher sometimes. Moreover, in the name of political discussion, there was a chance of participants blaming people who were from the community, which can in turn create chaos.

Studies report that the positionality of the researcher such as gender, class and race affect the research process from the articulation of research questions to the analysis and presentation of data (McCorkel & Kristen, 2003; Naples, 2003). One's social position influences the total gamut of research choices and influence becomes problematic when researchers occupy a privileged position in society.

Researcher's positionality as a faculty member of a public university in Bangladesh and outsider status, including gender and religion, affected this research process to some extent. The interviewer and researcher inherently have more power compared to the interviewee (Lee, 1997) in terms of knowledge on these particular research issues. The researcher avoided and asked research assistants to reduce the issue of positionality. The researcher considered that studying groups and respondents are experts (McCorkel & Kristen, 2003) in terms of knowledge and experiences on disaster, gender and vulnerability discourse compared tothe researcher.

It was extremely difficult to explain the purpose of this study. The people of Sutarkhali and Kali Nagor village obtained regular aid, assistance and relief from GO and NGOs. The researcher was asked multiple times how they will benefit from this research and was helpless in answering them. The researcher knew that it is extremely likely not to benefit them directly because they are benefitted by GO and NGOs' relief, aid and assistance. However, the researcher had to inform them that this study is writing for an academic purpose and to some extent for policy level. The way you (community people) suggest for your (community people) betterment, this study will write in the study report exactly the same. Finally, policy-makers, GO, NGOs and international agencies might consider your suggestions and accordingly they might take initiatives to help people of Sutarkhali and Kamarkhola union.

There was a geographical constraint of this study. Going to villages every morning, travelling 2–3 h and coming back from the village at the end of work every day was a big constraint. The *trawler* was used for travelling and some area was excluded because of inaccessibility via boat. In this study, only the area accessible via boat has been included. The areas excluded from this study may have different opinions and experiences compared to the area included in this study. Late morning and early afternoon are the only times the research team approached respondents and participants, which was an important time for community people to work in agricultural fields. Because of time constraints, some of this group might have been excluded from this study because they were only available late afternoon or in the evening. However, respondents in the survey study, FGDs and in-depth individual interviews had diverse backgrounds and education.

Farming time and busy schedule of people in Sutarkhali and Kali Nagor meant that the complete data collection took more than the anticipated time period. In particular, the researcher had trouble conducting FGDs with different committees such as UDMC, PIC, CPP and NGO activists because of their busy schedules and unavailability of all members. For example, NGO activists used to go to the villages in Dacope *upazila* on weekdays only. Some lived outside research areas and some travelled from Khulna (district headquarter) every day. They were unable to give time during office hours because they were busy with their job responsibilities.

Background, field of interest, and values of researcher influence the results of this study. For the researcher, many biases and pre-assumptions play an important part in the research process. Assumptions and biases are attributed to the professional background, academic background and future plans of the researcher. Researcher's previous experience, in addition to future plans with the expertise on this particular issue have affected this study in many ways. Researcher believes that socio-cultural and institutional practices in the traditional rural society in Bangladesh create disproportionate vulnerability in Bangladesh and in coastal Bangladesh. Moreover, the researcher believes that disproportionate vulnerability for women could be removed from society if women were rightly evaluated on their potentiality and if they achieve proper scope to utilise their merits. As a researcher, having such preconceived ideas because of previous experiences might limit the objectives and possibly over-optimistically impact of the study to some extent. As a researcher, this limitation kept researcher equally focused on community and household experience.

2.9.1 Ethical Considerations and Confidentiality

Ethical consideration and confidentiality arises at various stages in social research and role of values in research process becomes a concern (Bryman, 2008). The researcher took ethical issues and values as seriously as possible and attempted to give value to the issues of the community people in the research settings. However, steps were taken to ensure the participation of respondents and discussants with the freedom to withdraw sensitive information, voluntary participation, being informed well ahead, anonymity and confidentiality. Before starting the study, all participants were informed about the aims of the study and that all the information has been retained as the property of the researcher and must not be used for any other purpose except this study unless survey respondents, FGD discussants and individual interviewees gave consent. They were informed that the information sought from respondents and participants would be at a time convenient to them and that it must be on a voluntary basis.

Punch (2006) argued while the success of the study depends on the willingness of individuals to participate if at any time such participation places an individual at risk or causes undue stress, participation should not be pursued. Although relatively researcher has higher status because of job and higher income, he was dependent on the people of the community for access to data for this study. The participants and local community warmly accepted the research team, the trusted researcher and the research assistants. They helped as best as they could and in return the research team assured them that what goes in this research will not harm them in any manner.

Consent related to interviewing women was gained by arranging meetings with the community people because there are cultural barriers and restrictions for women to speak with outsiders. To obtain trustworthy relationships with community people and heads of households was extremely important because it influenced the women from the village in assisting with research work. The researcher's frequent acquaintance with them over a long time period helped build trust. The appointment of four female research assistants helped collect data from woman respondents and make the women feel free to answer questions.

The research notes and questionnaire were locked and protected by password and only the researcher had access to raw data. In the months when the researcher was conducting fieldwork, the participants and community people shared plenty of personal stories with the researcher but have omitted certain sensitive information, including illegitimate relations for the participants' well-being. For the researcher, the priority was not only research but betterment and interest of participants and respondents. On two occasions, the researcher travelled from Dhaka to Dacope to conduct FGDs with NGO and PIC committee; moreover, the meetings were cancelled at the last moment, and both FGDs were completed later.

2.10 Summary of the Chapter

The research design began with selecting the research sites, and these were purposively selected based on the research questions and research objectives. After finalising the research site, the immediate next step as a researcher was to form data collection tools for both quantitative and qualitative data for the triangulation of data and to maximise the reliability and validity of data. The pre-testing of data was followed by formulation of data collection tools. After finalising research tools, the researcher built a rapport with community people to collect data from the field the following steps such as data collection from the field, data cleaning, verifying, coding, recoding and data analysis. The schematic shows the complete process of the research design (Fig. 2.2).

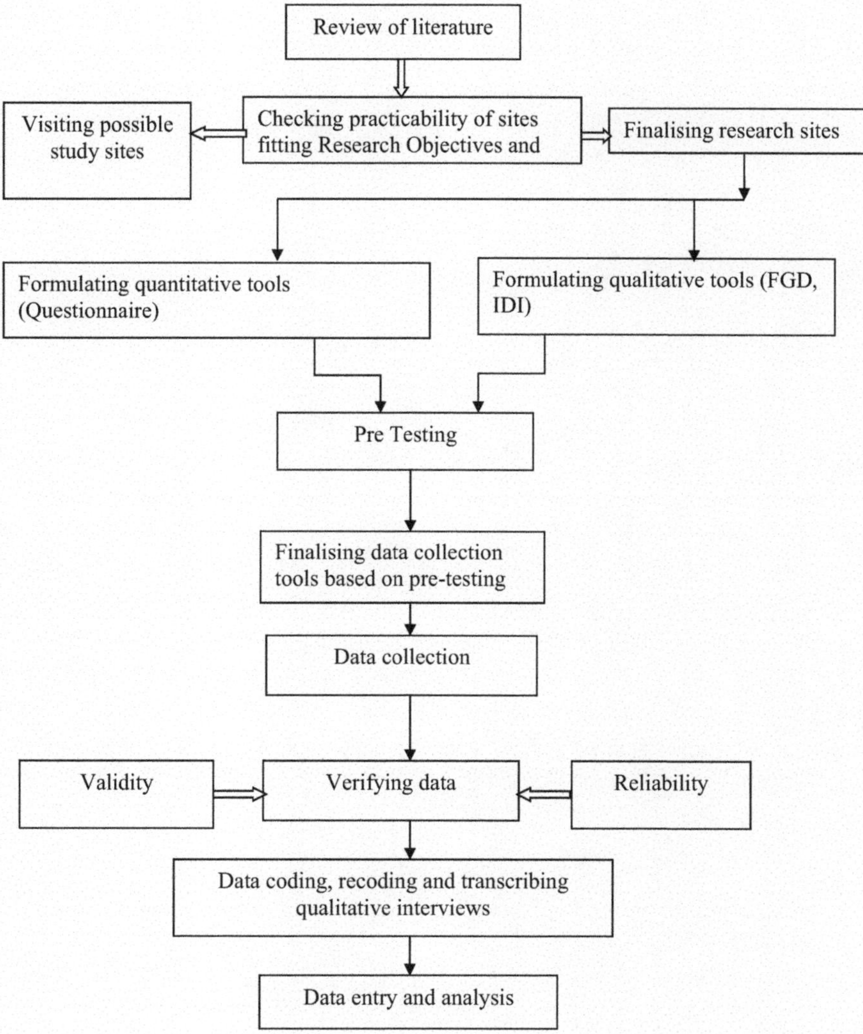

Fig. 2.2 Flow chart of data collection

Chapter 3
Disaster, Women and Vulnerability in Everyday Life

3.1 Introduction

During a disaster, women in the Bangladeshi coastal area are compromised by the lack of access to information, food, safe water, sanitation facilities and healthcare facilities. Using the descriptive analysis of quantitative data and quoting verbatim from the participants in the FGDs and in-depth interviews, this chapter address the experiences of disaster, vulnerability in shelter centres, health vulnerability of women, food scarcity and practices for taking meals, collection of fuel wood and water and vulnerability, relief/aid distribution and vulnerability, vulnerability of female-headed households, *purdha* practices, dissemination of warnings and vulnerability and problems of male migration. In addition to the vulnerability of women in everyday life, this book focuses on the coping mechanisms (which are often considered as immediate emergency response after a shock) and adaptations that tend to relate to ways through which households change their actions and behaviours to be better equipped to deal with future shocks (O'Donnell & Wodon, 2015).

3.2 Infrastructural Facilities and Vulnerability of Women

Gender issues are not the manifestations of disaster-related crisis situations but are prevalent in society, operational and visible in daily life at the level of individual, family, community and society at large and institutionally reflected in social and cultural norms. Gender aspects in social and community organisations lead to substantial discriminatory differences in how men and women of all age groups experience and deal with disasters (Ariyabandu, 2009; Fothergill, 1998). The results of the study explore that shelter centres lack adequate services, including a cooking place, community clinic, electricity and water reservoir, and the atmosphere in the shelter during a disastrous is abnormal. Vulnerability and discrimination of women

M. M. Khan, *Disaster and Gender in Coastal Bangladesh*,
https://doi.org/10.1007/978-981-19-3284-7_3

Table 3.1 Problems identified by the respondents in shelter centres

Facing problems in shelter centres	N	%
Yes	218	82.9
No	45	17.1
Problems in shelter centres (multiple)		
Food crisis	172	65.4
Drinking water crisis	167	63.5
Shower problem	144	54.7
Sanitation problem	143	54.4
Fuel wood crisis	119	45.2
Health problem	105	11.2
Purdha maintaining problem	58	22.1
Sexual harassment	10	3.8
Theft	10	3.8
Social restrictions	3	1.1
Others	3	1.1

Source Fieldwork 2019

can be seen more while the normal functions of society are disrupted and catastrophic disaster disrupts social structure for a long time because of the lack of change according to respondents.

In this study, 83% of respondents (Table 3.1) agree that they face problems in shelter centres and these are very much related to inadequate services. Sanitation problems (54%) and shower problems (55%) are identified by the majority of survey respondents as major problems along with others such as cooking place problems, food scarcity, sexual harassment and social customs. The problems are multiple, e.g., because of the lack of a cooking place, water and fuel wood, women face problems as they are obliged to be responsible and they cook under the open sky.

Safety is an important concern for women during and after a disaster because their reputation and the family honour are dependent on their actions; however, problems in the shelter centre, including the *purdha* system, lack of separate and sufficient sanitation, lack of shower facilities, insufficient accommodation, overcrowding, and lack of a separate room for women and disabled people disrupt the safety and security of women. Moreover, women's mobility is significantly restricted by practices such as inter-gender interaction, risk of sexual abuse, social condemnation and clothing. When women get into shelters, they are compromised by the crowded conditions, no separation between sleeping spaces, and no designated spaces for lactating mothers and pregnant women. Young girls are humiliated and endangered by the lack of space and privacy (Ahmed et al., 2008; Alston, 2015; Cannon, 2008; Nasreen, 2008; Rashid & Michaud, 2000; Save the Children, 2006). The relevance and importance of results regarding problems of women in the shelter centre is reflected by one vulnerable woman as follows:

> The shelter centre cannot accommodate too many people and there is no separate place for women. Eighty women took shelter from their para to the shelter centre and one of those eighty was pregnant and she was nearly due. The pregnant woman gave birth to a child in the miserable atmosphere of shelter centre. We created an artificial curtain hanging saree surrounding the spot. Taking care of the new-born baby caused considerable suffering and there was no privacy for the new-born baby's mother in the shelter centre. (Woman, 60, class V, FGD with female-headed household)

In terms of problems faced by women in shelter centres, a male respondent opines:

> Pregnant women had to face the worst sufferings. They cannot walk and freely move. Pregnant women require a safer and particular place but there is no separate and safer place for pregnant women. Men can reside in the shelter centre but it is really difficult for pregnant women to stay in such a crowded place. (Male, 50, class X, FGD with PIC)

Access to sanitary latrines is difficult during a climate crisis, and extremely poor and unhygienic sanitary conditions in shelters and homes result in urinary infections, diarrhoea, dysentery, skin diseases and hygienic issues in the study area; these limitations are recorded in other studies (Ahmed et al., 2008; Alston, 2015; Nasreen, 2008, 2010; Rashid & Michaud, 2000). The respondents report that, although the lack of separate toilets, kitchen and sleeping room in older shelter centres (Fig. 3.1, picture 2) create vulnerability for women, these conditions and facilities have been improved and addressed in the newly designed and built shelter centre (Fig. 3.1, picture 1). A woman UP member reports the following problems in the shelter centre:

> There is no end to the number of problems faced here. I cannot even finish listing them, and there is not only one problem rather there are multiple problems, e.g., sleeping, prayers, shower, toilet, and cooking problem. (Woman, 38, class XII, FGD with UDMC)

Women share a common set of sanitation facilities with men, and they change their wet clothes in common areas and, to some extent, in front of men because they can only create an inadequate artificial curtain. The respondents report that in the shelter centre, women, particularly young women, newly married wives and children, are extremely insecure. Moreover, women possibly suffer from increased mental strain and bear the brunt of certain social constraints, e.g., they are ashamed

Fig. 3.1 Shelter centre at the study area

in using public latrines or being seen by men when in wet clothes. Consequently, the feeling of insecurity, sexual harassment and criminal activities that originates from using common and unprotected places in crowded shelter centres increased after a disaster (Rashid & Michaud, 2000; Saha, 2015). The male respondents agree with the females while conducting FGD with UDMC as follows:

> Women cannot move easily. They cannot take shower and change clothes and the environment is too bad for young women. We had been living there taking dry food because there is no arrangement for cooking. We faced problems in using the toilet. (Male, 33, class IX, FGD with UDMC)

The respondents report that after the cyclone *Aila*, most people took shelter on the *beribadh* under the open sky and developed a shanty, a temporary house. People from different stages of society then took shelter under the open sky. Living under the open sky acted as a threat to the security of women and young girls. Initially, there were no toilets and sometimes the toilets were in an open place (Fig. 3.2, picture 1) and away from the residence (Fig. 3.2, picture 2); therefore, they did their bathroom activities at night in an open area. Husband, wife, young children and all other family members were living in the same room. Sometimes, miscreants disturbed them with torchlight along with the unwelcome touching of women's bodies at night. Women had a miserable time but they hid their problems and did not disclose them to anyone. The experience of living under the open sky and sexual violence against women is described below by certain respondents:

> Miscreants and bad boys of the society disturb women, young women, and girls. During a disaster, when women take shelter under an open sky and in the shelter centre, miscreants and bad boys get a more favourable atmosphere for sexual harassment. Under an open sky, women cannot hide their activities. They change clothes and take a shower in an unsecured and open space, which in turn provokes miscreants and bad boys. Moreover, sometimes an illegitimate relationship develops between a man and women because they are staying in a place for long time. (Woman, 33, class X, FGD with vulnerable women)

Vulnerable women are subject to sexual violence, including rape; illegitimate physical relations with consent and offers for sexual relations in lieu of giving advantages; physical abuse; unsolicited or unwanted body touching; molestation; and a

Fig. 3.2 Vulnerable toilet and distant location from the residence

sense of insecurity or fear. Survey participants and focus group respondents report that sexual violence, sexual harassment and offers for extramarital relations, and illegitimate physical relations are prevalent in Sutarkhali and Kali Nagor. Furthermore, these problems are exacerbated during a disaster compared to normal times.

This study reports that, during a disaster, women sometimes lived alone with their children. Certain respondents report that, after *Aila*, husbands deserted their wives and children; sometimes, they moved away or migrated to the city or other districts and started another family by getting married, which is not legal. In such a chaotic, unorganised and unplanned place, women and girls were exposed in the study area to sexual violence and insecurity. Sayeed (2010) argued that when other people or a male neighbour knew that a woman did not have a male protector, she would be open to abuse. In terms of chaotic conditions during disastrous times and for sexual harassment, Fisher (2010) argued that lack of security and chaotic conditions created an environment in which girls and women were extremely vulnerable to violence, although women were perhaps most vulnerable to sexual violence in the first few chaotic weeks of the disaster.

This study demonstrates that respondents and discussants in Sutarkhali and Kali Nagor were reluctant to share information about sexual harassment and sexual violence whether they had these experiences or whether they noticed others during a disaster. However, 39% of respondents shared that they felt insecure and had experiences of sexual harassment. Respondents report that they are used to and accept the chaotic atmosphere during a disaster. The scenario of sexual violence and insecurity of coastal area is reflected by a male discussant as follows:

> Economic insecurity of coastal people is being solved by GOs and NGOs by considering different programmes such as relief and food for work. However, problems such as sexual harassment, sexual violence, illegitimate relations, extramarital relations, and eve teasing are not being totally solved, although such incidents are being reduced. Often, these incidents do not happen openly and victims do not disclose the incident. (Male, 45, class 16, FGD with PIC members)

A vulnerable woman from female-headed households reports sexual violence and insecurity as follows:

> During a disaster, there is a chaotic atmosphere in the locality, particularly at the shelter centre. There is scarcity of everything, which creates a favourable environment for bokhate young men and teenagers to commit crimes and trap women, particularly the destitute women, by offering different advantages. (Woman, 45, illiterate, FGD with female-headed households)

While sharing the opinions of women with UDMC members during a FGD, almost all male members disagreed with the opinion of female members. UDMC male members report that they have never seen and not even had reported to them any incident related to sexual harassment and violence during a disaster. However, in terms of sexual violence in the study area, an NGO activist reports the following:

> Indeed, there is sexual harassment, sexual violence, and illegal physical relations in the coastal area during a disaster. Widows and single woman are the primary victims and vulnerable but these incidents take place secretly. Local leaders and influential persons mostly have

> the opportunities to exploit the weak. Moreover, UP members will not confess these incidents because, in most cases, either they are directly involved or people who are close with UP members are involved with sexual harassment. (Male, 42, class XVI, FGD with NGO activist)

However, not everyone sees the post-disaster situation to be worse in every respect. In terms of the status of gender-based violence in the study area in the post *Aila* period, the elected union chairman reports the following:

> Gender violence, husband-wife chaos and dispute, sexual harassment, and other issues and problems have decreased in this area. As a chairman, I was used to doing shalish almost every day but now I do very few. I believe that the number of incidents has reduced because of letting women work and earning along with a male member. I cannot say there is no gender-based violence but it has reduced. (Male, 49, class X, FGD with UDMC members)

The chaotic atmosphere along with economic hardship creates favourable conditions for sexual harassment of women. Economic marginalisation and lack of access to resources left women dependent on men and vulnerable to manipulation and sexual exploitation (Fisher, 2010). An NGO male activist reports the following:

> Some women may have to get engaged in illegal physical relationship and many of such relief getters are widow or destitute helpless women. UP chairman and members took this opportunity. They lure destitute women by offering a relief card, widow card, and old aged allowance card and then offer them to enlist their name for different aid and assistances provided by GO. Because of scarcity and need, women become compelled to do as they are asked. (Male, 44, class XVI, FGD with NGO activist)

To cope with problems of sexual harassment, sexual violence and insecurity, women had to take certain measures. In the coastal study area, the most common coping mechanisms against sexual violence identified by the respondents include village arbitration, which is led by a *mattabbor* who keeps sticks handy and leaving lights on at night, shares with close relatives, and takes help from NGOs (legal aid/service, shelter support). However, certain respondents report they do not take any action (they hide these incidents) because they do not feel secure about sharing incidents with their husbands because thee husbands are used to blaming women, which worsens the relationship between husband and wife. An NGO activist male discussant portrays the following coping mechanisms (Fig. 3.3):

> A few days ago, we heard about an incident of sexual harassment. We helped the victim and brought out a procession and demonstrated in front of the UP against the criminal. Later on, the criminal managed the victim and solved their problems by giving money to the victim's family. It is difficult for victims to publicise their problems. The problem is locally solved; in most cases, it is mediated by local leaders and not publicly disclosed. (Male, 48, class XIV, FGD with NGO activists)

Another male participant describes in a focus group with PIC:

> There are certain cases of sexual harassment, including rape, in the locality. However, sometimes, the relatives and family of the victim do not want to file the case against the offender because they are concerned that the offender would create propaganda out of the incident, which could be disgraceful for the victim's family. Therefore, the incidents remain under cover. (Male, 50, class XIV, FGD with PIC members)

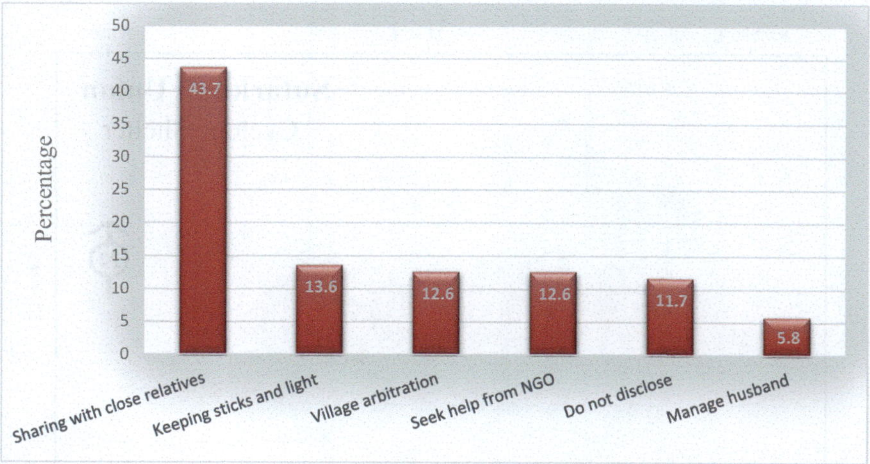

Fig. 3.3 Coping mechanisms against sexual harassment

Survey respondents and focus group participants report that the large tidal waves of cyclone *Aila* demolished the *wabda*. The local people did not have sufficient time to make decisions and they just went straight to the shelter centre. Immediately shifting to the shelter centre was more problematic for women than men because of the location of the shelter centre. In terms of the distance of the shelter centre, a woman discussant describes (Maps 3.1 and 3.2):

> There are not enough shelter centres in the locality; all shelter centres are far away and the communication is very backward. There is no place in the shelter centre even to stand. You cannot even find a place for sleeping. The only way to get into the shelter centre is to walk. The distant location of the shelter centre and backward communications creates problems for women. (Woman, 30, Class IX, FGD with vulnerable women)

In terms of transport communication problem, vulnerability and shelter centre, a male UP member reports the following:

> There are insufficient shelter centres in this area. The most important limitation is transport. Because of bad transport and physical weakness, women and particularly pregnant and aged women could not quickly move to a shelter centre. Children and old people are then treated as burdens. (Male, 42, class X, FGD with UDMC)

3.3 Disaster and Health Vulnerability of Women

Discussants and respondents report that year-round diseases are widespread at the Sutarkhali and Kali Nagor villages because of the geographic location of these villages, low incomes of people in this community and greater reliance on and contact of villagers with polluted and saline water. Extreme natural events such as the cyclone *Aila* severely affected the health of people in the coastal community. A high majority

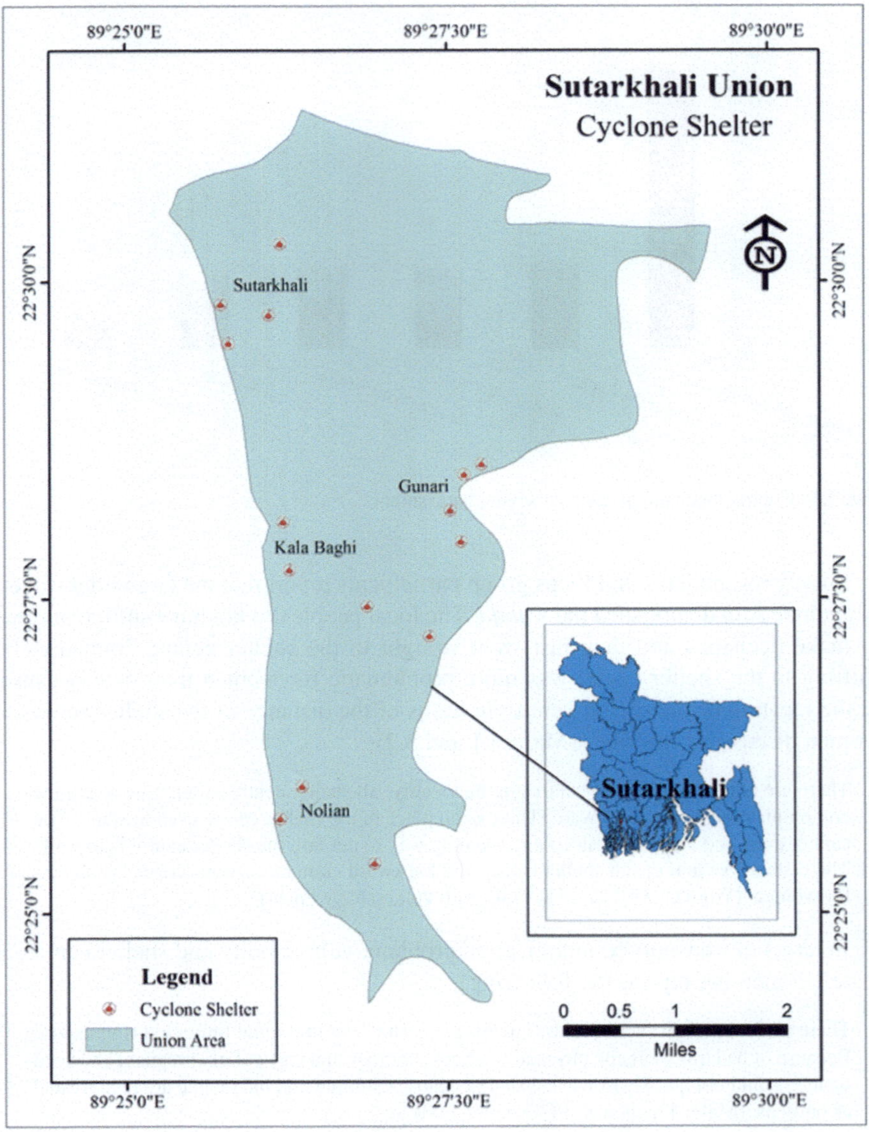

Map 3.1 Location of the cyclone shelters at Sutarkhali Union

(83%) of respondents agree that they faced diseases, health-related problems and problems in getting treatment. The most common diseases in the coastal area during a disaster are cholera, viral fever, jaundice, skin diseases and malnutrition related diseases. Cholera and skin diseases easily spread in the coastal area because of the scarcity of drinking and cooking water, the presence of polluted water and increased density of saline water, as described by the respondents. Survey respondents and

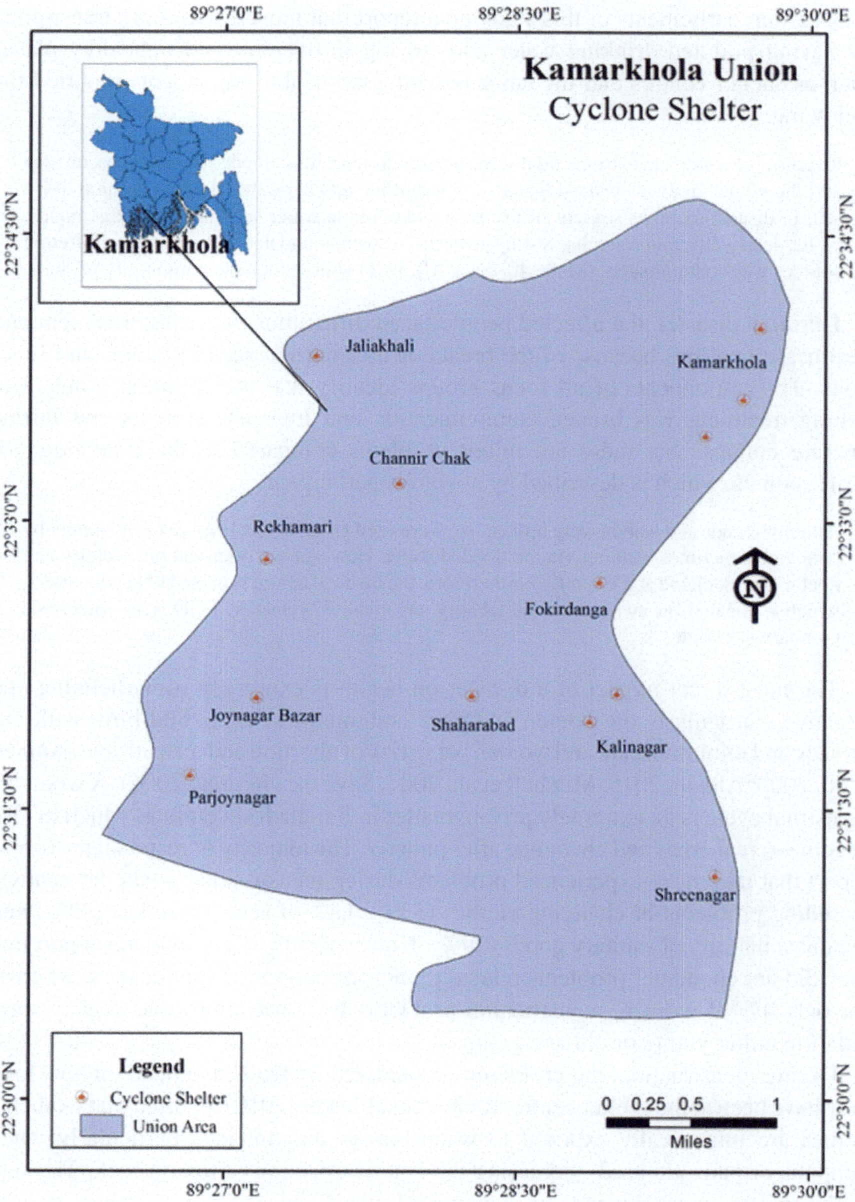

Map 3.2 Location of the cyclone shelters at Kamarkhola Union

focus group participants of this study area report that most diseases are transmitted by having polluted drinking water and staying in overcrowded unhealthy places such as shelter centres and the embankment. One of the respondents reported the following:

> Overflow of water and extreme tidal surge during cyclone Aila, flooded the total coastal area and the whole area was water-logged for a long time and agricultural lands became like a sea. In disastrous times sources of drinking and cooking water such as tube wells, ponds, wells, deep wells, rain water harvesting projects etc. were inundated and water from different sources was contaminated. (Male, 40, class XII, FGD with vulnerable community people)

During a disaster, the affected people faced difficulties in getting treatment and healthcare facilities because of the breakdown of normal social systems and functions. The respondents of all focus groups identify that the common problem in getting treatment was broken communication and transport systems and infrastructure collapse but these are minor problems compared to the availability of professionals, which is described by a woman participant:

> Although roads and wabda were broken, we were used go to medical centre for treatment by boat and sometimes walking via the flooded water. However we often did not identify any doctor and medicine at the centre. Furthermore, the medical centre is in the UP headquarters, which is located far away from the locality. (Woman, 37, class V, FGD with vulnerable community people)

The most direct impact of a disaster on health is extremely overwhelming and negative, particularly on women and poor communities, e.g., child birth with our privacy and minimum care and women being risk of abortion and miscarriage (Ahmed et al., 2008; Alston, 2015; Mitchell et al., 2007; Save the children, 2006). A woman's menstrual cycle is an extremely private matter in Bangladeshi culture, which in turn becomes a real crisis and challenge after puberty. The majority of respondents (66%) report that they have experienced problems during menstruation in shelter centres, including problems of changing clothes (53%), lack of separate toilets (44%) and the unavailability of sanitary goods (29%). However, 30% of respondents report that they did not encounter problems related to menstruation in shelter centres, whereas for only 4% of women, menstruation problems were not applicable as they were either in either young or old age group.

During menstruation, the problems encountered by women depend on how long they have been at the shelter centre. Richter and Flower (2010) reported that women's bodies are intrinsically exposed to water-borne contaminants, particularly when tampons or pads are used, increasing the risk of infection or toxic shock. The stay at the shelter centre depends on how quickly the flood water goes away from the locality. In particular, after *Aila*, certain people stayed at shelter centres for certain months. Problems encountered by women during menstruation are reflected in the following statement:

> During menstruation period, women suffer a lot because of a lack of proper sanitation. Women cannot wash their menstruation rags; they are to stay in wet dresses, because they cannot make them dry. The experiences for women particularly for young girls become severe at the disaster shelter centre. In such a condition, women and girls are unsafe even

from neighbouring men or close relatives. Many women and girls do not go to a shelter centre because of the lack of security and unhealthy sanitation facilities. (Woman, 22, 14 year schooling, FGD with vulnerable women)

Respondents describe that after returning back home; they reported a severe drinking and cooking water crisis; because of scarcity of drinking water, they sometimes ate raw vegetables. Sometimes, women from different households became victims of the anger of male members because women are responsible for cooking and males do not want excuses. The lack of drinking water pushed women to use polluted water from the pond and surface, which increased the potential for diseases. Both surface and ground water used for drinking and cooking purposes became unsuitable for human consumption over a large area of the South-western Region of Bangladesh (Roy, 2004). Because a large section of the population is not served by treated water supplies, they are often exposed to and affected by diarrheal diseases.

Furthermore, if someone does become sick in the family at the time of disaster, it is women who take responsibility more on their shoulders compared to men to look after the sick. This study reports that poorer women take risks and collect medicinal herbs from distant places during the floods despite the *purdha*. Men, however, rarely do such work because providing herbal treatment or nursing is not their gender-assigned role, which is consistent with another study (Nasreen, 2012). One male describes treatment and health problems of the study area as follows:

People in coastal disaster-prone area are not conscious in terms of health problems and treatment because the majority of the people of the coastal area are poor and illiterate. For example, most husbands keep their wives at home for normal delivery/birth of the baby because they believe that caesarean for baby is anti-Islamic. Therefore, in most cases, local unskilled dima help at the birth of the children. Kabiraji treatment, local quack, and herbal treatment is the major treatment for people in coastal areas. (Male, 47, class XIV, in-depth individual interview)

Another woman describes the herbal and *kabiraji* treatment as follows:

I do not think there is any modern treatment in this area. People primarily depend on herbal, traditional kabiraji treatment, and homeopathy treatment. In most cases, a woman of the family manages herbal care for a sick member of the family because women know about leaves and herbs better. (Woman, 23, class X, in-depth individual interview)

To some extent, if these poorer families get modern treatment, it is done by selling their valuable assets including the women's gold ornaments. However, if poorer families lose their assets such as ornaments, livestock, poultry and small areas of land, their economic condition deteriorates and the burden for this heavily falls on the shoulders of women. Therefore, most of the poorer women, despite their own sickness and other responsibilities, try their best to cure sick family members to avoid the expense of modern treatment. Regarding treatment a male member describes the following:

People in coastal areas focus on and value the opinion of the senior family members and society for treatment. Moreover, we are Muslim, we have not left our religion and cannot reject the Quran. Many religious leaders give a tabij for different diseases and people believe in that. Sometimes, it works and sometimes it may not work but the tabij from a religious leader is a very common and frequently used traditional treatment. (Male, 47, class XIV, in-depth individual interview)

Saline water in coastal areas became more problematic for women because of the nature of their activities. Studies report that gynaecological problems, menstrual pain and water-borne illness are expected to be more widespread among nutritionally disadvantaged women (Ahmed et al., 2008; Cannon, 2002) in saline water-affected areas. This study shows that women are in touch with polluted and saline water for cooking, cleaning, caring for livestock and poultry and household work almost all the time. Even if they want to avoid polluted water, they cannot do that because of the location of their homes and lack of alternatives (Fig. 3.4, picture 1 and picture 2). The following statement reflects the health vulnerability of women because of saline water:

We were infected by skin diseases when dealing with saline water all or almost every time; it causes irregular menstruation and menstruation-related disease too. They sought treatment to get rid of problems and some of them got relieved from such problems but some others are still suffering from them. (Woman, 31, illiterate, FGD with female-headed households)

Do you think health vulnerability will be lesser if there is a community clinic (a community clinic is a small clinic at the grassroots level present in the remotest areas for health education, health promotion, treatment for minor ailments and identification of emergency and complicated cases for an effective referral linkage and higher facilities for better management; http://www.communityclinic.gov.bd) in the locality other than the union health complex at the UP headquarters. In response, a vulnerable woman describes:

There is a community clinic; it is just a building standing over there and is a clinic just by name. We did not see any doctor to come in the community clinic and treat community people, although doctors are regularly supposed to be there. Doctors stay at the Upazila Sadar and come once in a month to draw their salary and sign for the whole month. (Woman, 45, class XII, in-depth individual interview)

Fig. 3.4 Using polluted water

This study shows that coastal people are not getting proper treatment for diseases and health-related problems. Regarding additional problems, a woman describes:

> Women of this locality keep to themselves; they do even remain quiet over their sickness and do not raise voice because of their illiteracy. Finally, backward transportation, unavailability of a vehicle, and unavailability of doctors create additional problems for women's health. (Woman, 39, class IX, in-depth individual interview)

The respondents of this study show that insufficient medicine and unavailability of doctors (60%) during a disaster caused more problems particularly for treating women. Even when doctors stay in the union health complex and visit patients, they do not carefully listen to the patients' problems and do not listen to details about health problems, which have been termed by respondents as ignorance and negligence (13%) to patients by doctors. Problems mount up while women are the only earning member in the family; under these circumstances, women cannot even visit a doctor for treatment and health care. Note that 15% of women comment that despite sickness they had to perform household activities and most families are poor and female-headed households. In addition to these women, 34% of respondents report that scarcity and shortage of medicines and distant locations of medical centres are limitations for coastal women to get health care and treatment. Although it is small in number, 11 (4.6%) women report that they were unable to visit a doctor for treatment because of ignorance by husbands and family members. A vulnerable woman then describes:

> I got really sick after Aila. I could not even stand and walk, and I was laid on bed for a month without any treatment and medicine. However, I could not go to see the doctor because my husband migrated to Gopalgong for work and asked me not to go to the doctor with any other man. My husband then took me to the doctor when he came back. (Woman, 41, illiterate, FGD with vulnerable women)

However, often for the case of severe disease and health risk, people in the study area did not follow cultural barriers and felt compelled to take patients to doctors and did not wait for a male's decision. The respondents report that they attempted their best during a disaster to keep family members safe and free from disease despite the unhealthy atmosphere, close proximity and polluted water. They used their own knowledge and traditional skills to avoid using contaminated water and keep their family safe from disease. Survey results demonstrate that certain respondents took *kabiraji* treatment (9%), religion-based treatment (6.5%) and treatment by village doctor (6.5%), which includes treatment with different leaves and grass and sometimes wearing a *tabij* around necks or arms, *ghar bhanda*, and pouring water over the heads, thus administering juice of leaves and grass to stop bleeding and covering patients with heavy clothes to stop shaking and get rid of high temperatures. A woman notes the treatment of coastal area as follows:

> People in the south area primarily go to kabiraj and get herbal treatment. They do not think even of modern treatment or go for scientific treatment; rather, they depend on leaves and herbs. (Woman, 39, UP member candidate, class 9, in-depth individual interview)

Note that 36% say that they went to GO doctors at the *union shastha kendra*, and 7% of respondents went to NGO doctors for treatment despite the unavailability and uncertainty of whether they would see the doctor. Because of monetary problems, poorer women depended on visiting GO doctors in the union health complex despite uncertainty and NGO doctors who only occasionally came; sometimes only once a year. NGO doctors frequently visited coastal area during a disaster but the frequency of visits and treatment by NGO doctors reduced as the situation normalised. Sometimes (almost every year), GO and NGO doctors established a treatment camp in the community for two or three days once or twice a year; moreover, the poor people waited for six months and sometimes a year to get treatment from the camp doctor.

Survey data show that of the 263 respondents, 19% reported that they did not take any treatment; however, 12% reported that they took self-treatment and nursed themselves during illness using their own knowledge and experience. Note that only 4% report that they could visit a private doctor and clinic in the *upazila* and district city for modern and better treatment. The discussants report that the number of respondents who sought better modern treatment is extremely low in the coastal area because of distance, cost and bad communications. A woman describes:

> It is true that there are no modern treatment facilities for coastal community people but there is quack/traditional doctor (a local doctor with self-trained or poorly trained) and village doctor. People do not even care to go to village doctor. They do their own treatment, e.g., coastal people hang banana leaves for pain at neck or drink hot water with lemon and salt. (Woman, 39, class IX, in-depth individual interview)

3.4 Food Scarcity, Practices About Eating, Processing Food and Women

During a disaster, while affected by cyclones, poor people are less able to rely on their savings and possibly reduce their food intake (O'Donnell & Wodon, 2015) and the role of women is to be shock absorbers (Thieme & Seigmann, 2010). For example, the reduction in food intake by women has been defined as *shoye jaowa*. At Sutarkhali and Kali Nagor, food security was completely shattered after cyclone *Aila* primarily because of the collapse of agriculture and land-based production systems because of the submergence of farm land under saline water and non-availability of land-based productive systems (Ahmed et al., 2007). Moreover, the purchasing capacity of coastal people at that time was shattered because of the scarcity of alternative occupations, including fishing and day labouring in shrimp *gher*. Purchasing food items requires the sustained flow of income, the latter being extremely hard to come by for poor families as demonstrated in this study.

The local economy was extremely stressed during this disastrous time such that no employment opportunities were generated in the coastal area. The results of this study show that cyclone *Aila* drastically shattered the livelihood options, including shrimp *gher*, fishing in river and sea, collecting fruits, honey, fuel wood, *gholpata* and driving boats; moreover, it resulted in an extreme financial crisis (Saha, 2015).

The impacts of *Aila* on occupations in the study area are described by a respondent as follows:

> Most occupational sectors were ruined by Aila. Crops in the field were damaged, and farmers could not start farming land for a long time after Aila because of salinity and water logging. No fishes in the river and fisherman left his occupation. People who were dependent on jungle (sundarbans) for their livelihood did not visit a jungle because it was totally destroyed by Aila. The consequence of devastation of Aila and damages of earning opportunities created food scarcity and limited the access to food as people did not have any income. (Male, 75, illiterate, FGD with vulnerable community people)

In this study, 98% of respondents report that they had food scarcity during disastrous times. Climate-induced disasters and food insecurity affect both women and men; however, the burden of coping heavily falls on women's shoulders in Bangladesh (Nasreen, 2008). The respondents' report that access to food in the family varied as per socio-economic background and the economic contribution of respondents. Survey participants and focus group discussants report that it was women who had more responsibility to cope with food scarcity and had to adjust their consumption habits. Women and girls were used to being served at the end of the food distribution at the table compared to male members of the family. In terms of having less food and being served last, a woman describes:

> After Aila, coastal people are in a more distressed condition and scarcity is more acute than before. The land has lost its fertility and production is only once a year. We cannot eat as per our demand. Therefore, women manage food scarcity by having less and last. (Woman, 39, class IX, in-depth individual interview)

While families receive an equitable share of food aid and assistance by different organisations, food is not equitably distributed among family members. Both women and girls are used to having less food and to being the last during food distribution and last to eat compared to male family members as stated by discussants. Sex and age biases were both the determining factors in food distribution revealed by Sen (1988). The respondents state that cyclone *Aila* caused food scarcity, and people in the coastal area had limited access to food and consumption because of crop damage (51%), unemployment (45%), lack of cash (68%) and unavailability of food (53%) in the study area during a disaster. The respondents report that price hikes of daily essentials, damage to crops, damages of shrimp *gher* and intrusion of saline water adversely affected the employment opportunities in the study area; moreover, the lack of employment opportunities, limited access to food consumption and purchasing capacity. Losing their harvest and their backyard gardens, women ordinarily feed their husband and children first and then live on the left over food, even while breast feeding when they require more food.

Women's status and position are inferior in the economic structure of Bangladesh; they do not find any room in the estimate of national income, and their access to service and business is limited by traditional practices (Nasreen, 1995; Oxfam, 2011). Researchers (Agarwal, 1986; Chen et al., 1981; Chowdhury et al., 1981; Nasreen, 1995) report that there is gender bias in the household food allocation in many societies, and this gender differentiation is acute where there is food scarcity, which is reflected in this study. Note that 64% of women agreed that they do not take meals before their husband or other male family members because the discussants argue that it is culturally expected that women should sacrifice their food for the survival of their husbands and sons. However, in terms of women (34%) who took a meal before their husband or other male members, most of them are heads of households or they are the senior members, or are living with a son. A woman and a man now describe the impact of gender status for distributing and serving food in the study area (Fig. 3.5):

> Normally, women believe that a man works outside and earns money and is contributing to maintain his family. Women maintain the better portion of the food for male members because they need to be physically strong and healthy. To keep male members healthy, a woman sometimes gives a bigger and two fish rather than one and it does not matter whether anything is left for the woman. (Woman, 23, class XI, in-depth individual interview)

> Men will eat food before women. It becomes a rule in our society. The rural coastal areas are a patriarchal, male-dominated society. Women always take food later and we have observed this practice since our childhood. (Male, 47, class XIV, in-depth individual interview)

In terms of the priority of serving food and meals, 58% of men and 31% of children were served first compared to only 1% of women. Other studies report that men in a Bengali society are given priority in food intake to ensure the continuity of the male line. If food is limited and short, the expectation is that women will eat less (Alston, 2015; O'Donnell & Wodon, 2015). The differences and unequal distribution of food occur for three reasons: cultural, extent of resources and functional arrangements, which are reflected in this study (Wheeler & Abdullah, 1988; Greenough, 1982 as cited in Nasreen 1995). The respondents report that patriarchal and religious obligation (76%), male's status as the earning member (15%) and (7%) men own the resources (7%), thus ensuring priority in taking meals by men. A woman describes cultural practice of having food by women as follows:

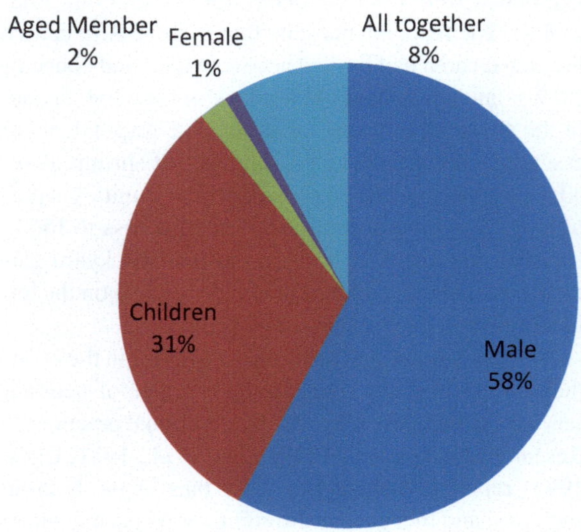

Fig. 3.5 Distribution according to priority in serving meal by gender

> Women should not and women do not eat before male members. If there is no food, it is better for women to starve rather than take food before male members. Women will eat only after serving males. Can a male be starved? No, it is not done. It is respect and community practice that women have to serve first and will take food later and last. (Woman, 29, class IX, in-depth individual interview)

Many respondents of the survey report that they have adjusted by taking less food (14%) compared to necessity and by first serving male members and children (11%). During a disaster, because of shortage and food scarcity, women sacrifice more although women are closely related to family food distribution and women are the food procurers in households. Ali (1984 as cited in Nasreen, 1995) reports that, for Bangladesh and India, during any type of food crisis when men cannot earn enough to feed the whole family, women become food procurers for household members and engage in other work for their family's survival.

Women's poorer nutritional status reduces their capacity to cope with hazardous effects (Alston, 2015; Cannon, 2002; Delica, 1998; O'Donnell & Wodon, 2015). Ninno et al. (2001) reports that in Bangladesh women of all ages are more calorie-deficient than men, and the prevalence of chronic energy deficiency among women is extremely high. Consequently, girls and women may eat last and least and can be malnourished long before the drought reduces the family's access to food (Enarson, 2000), which is depicted by a woman as follows:

> Although I am cooking food, I feel that my husband is working hard in the field. From that belief, I keep some extra food for my husband and take less or sometimes I do not eat. My husband does not care whether I have eaten or not, but I care about his health. Because I am not taking as much food as my husband, I suffer from malnutrition and different diseases. Moreover, because of nutritional deficiency, the growth of children is hindered. (Woman, 23, class XI, in-depth individual interview)

Women's own coping techniques and initiatives have become crucial for their family sustenance and ensuring food security because women are better at practising household management from a young age. This study reports that most households had no stable source of food during a disaster and most of them (64%) depended on relief for their food immediately after cyclone *Aila*, which is consistent with the results of this study (Saha, 2015).

The affected people in the study area have become dependent on other external assistance, including VGF, *Kabikha* and *Mati Kata*, of GOs and NGOs. Women played a role in managing food for other family members because women got priority from GOs and NGOs to be listed for relief and aid; moreover, women had more and easier access to relief and external assistance. Survey participants and focus group respondents argue that there was always seasonal food scarcity and shortage in their locality; however, disaster makes the scarcity of food extreme. In terms of their dependence on external sources during a disastrous, one of the vulnerable women describes:

> I am old and cannot work. My sons do not look after me. I am living in a jhupri (house made by grass and straws) on the wabda (dam), and I am dependent on relief and monthly old aged allowance of the GO and help from my neighbour. Sometimes, I eat once in a day or I just starve. (Woman, 80, illiterate, FGD with female-headed households)

Borrowing from friends and relatives is an extremely common coping mechanism during food scarcity, which is reflected in other studies such as O'Donnell and Wodon (2015). Women in the study area work outside (38%) and borrow from neighbours to overcome scarcity (35%) because at the end of the day women have to get food ready and are adjusting the household food consumption, which is a culturally established practice in the study area.

Both in normal and in disastrous times, women are primarily responsible for arranging/preparing food in Sutarkhali and Kali Nagor, and the respondents report that women's efforts are useful in food arrangement/preparation for family members. The survey respondents and focus group respondents report that whatever the amount available at home, women would separately store some food every day, particularly rice, to use in case of emergency; moreover, men were always ignorant about storage. This was uncommon for destitute people in the coastal area. In this regard, one woman reports:

> I do not have food even to eat once in a day; some days, I stay without having food, so then how can I save some. (Woman, 65, class IX, FGD with female-headed households)

Because women were primarily responsible in the coastal area for arranging food, one of the most common strategies to cope with food scarcity was sale of their assets. Respondents and participants were asked whether they sold any assets or. In response, 16% of women report that they have sold their assets, including jewellery, poultry, livestock and household goods. The study shows that the assets belonging to women and looked after by women, including jewellery and poultry, were sold during a disaster to help with food scarcity. In terms of selling jewellery to meet family requirement during a disastrous, a woman reports the following (Fig. 3.6):

> I received some jewellery from my parents. After Aila, my husband sold it because he needed some amount of money to migrate to Dhaka. By selling my jewellery, he took the money and gave some to manage the family. (Woman, 37, class VIII, FGD with vulnerable women)

3.5 Collection of Water, Fuel Wood and Vulnerability of Women

Gender enters all arrangements in Bangladesh; the household is identified as the core unit for analysing gender roles, relationships and gender division of labour by respondents. Household-related activities of women in the coastal community, including cooking, water collection, child rearing, and caring, are identified as "non-work" and non-productive because there is no earning from household work, which is treated as women's natural role, and men's work is defined as a market-based productive work.

> Males do not understand and do not even try to understand. Males live with their ego. Basically, the household activities of women have no importance and priority. Because of ego, male members are used to considering household activities are the activities of women

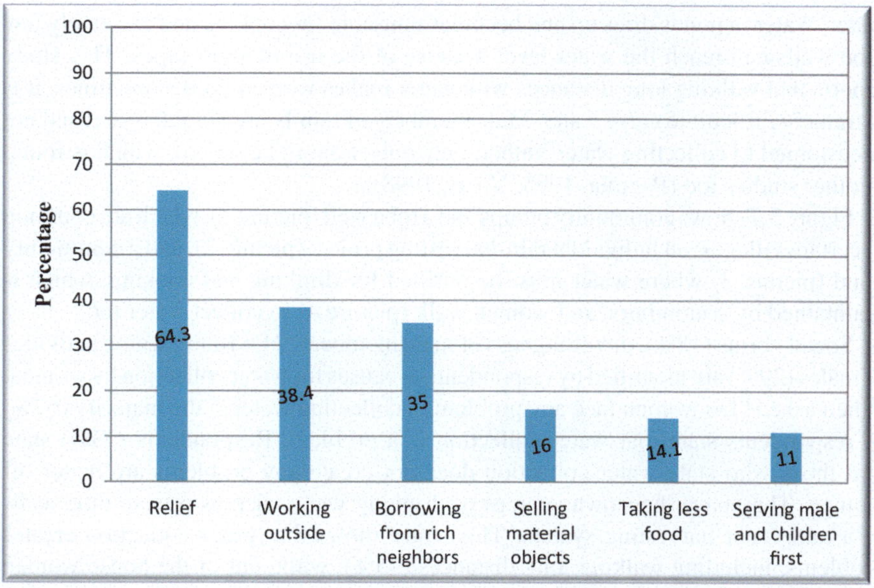

Fig. 3.6 Adaptive mechanisms to overcome food scarcity

> and non-monetary in nature. Therefore, a woman has to these activates because no money is being paid for household activities. (Woman, 23, class XI, in-depth individual interview)

In coastal areas, water collection is domestic work; however, women are not solely responsible for collecting water, although because of experience women know better about source and location of water and difficulties can be overcome. However, women are not receiving real appreciation for their contribution to the family. In Sutarkhali and Kali Nagor village, water sources were tube well (76%), ponds (10%), rain water (9%), relief water (66%), flood water (8%) and other. At both Sutarkhali and Kali Nagor, certain community ponds were maintained by the community for sweet water that was used for cooking and drinking. However, in addition to tube wells and community ponds, certain families had a container to store rain water (9%), which was termed as tanking water (rain-harvesting water mostly provided by NGOs), by the local people. When saline water entered the community water sources during a disaster, relief water supplied by GO and NGO became the major source of water (66%).

This study shows that water collection was normally evenly divided among groups: only women collected 33%, only men collected 35%, whereas both men and women collected 32%. In the early morning, women stood in the queue with other women to collect water from nearby sources. During a disaster, in addition to many other problems and responsibilities, collecting drinking water only became the responsibility for women, irrespective of their physical condition. Moreover, collection of water has become an acute problem in the dry season because of the depth of level of ground water and unavailability of water from other sources such as rain water. Water in ponds dries up and becomes unusable for cooking and extremely few tube wells can reach the water level because of the use of short pipes. This study reports that walking long distances with water makes women tired; sometimes, it is extremely difficult to carry water. Male members of family are not interested and not accustomed to collecting water; rather, they only want to be served, which is found in other studies too (Pagiola, 1995; Vinas, 1998).

Figure 3.7 shows community pumps and a tube well (picture 1), which are common in certain villages: an individual rain-harvesting project (picture 2), and a community pond (picture 3) where water must be purified for drinking and cooking, which is maintained by community, and women walk (picture 4) to collect water too.

Social norms (18%), unwillingness of male members (21%) and earning activities of males (29%) are identified by respondents as causes for water collection by women. When asked "Do women face any problem in collecting water?" the majority (67%) of respondents state that water collection is a problem. Respondents (33%) state that those who state water collection does not create any problems are better off women. They have their own sources of drinking water such as private tube wells or a rain water harvesting system. This study shows that water collection creates problems, including walking long distances (75%); while out of the house women suffered teasing (14%) and illegal offers (extramarital and sexual relations) (11%).

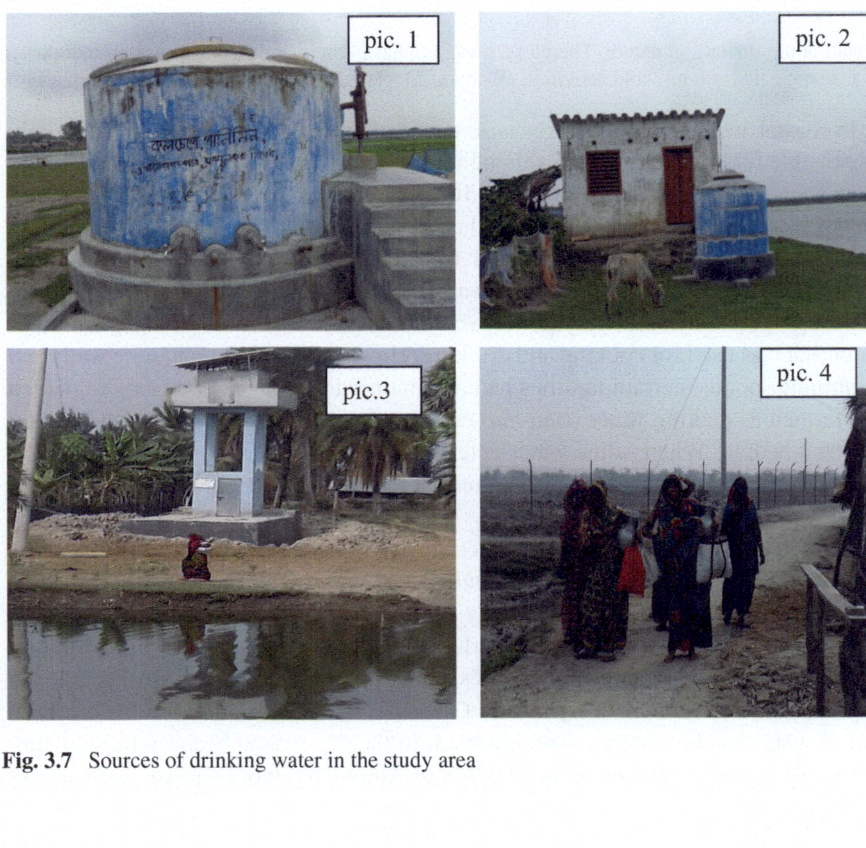

Fig. 3.7 Sources of drinking water in the study area

Collecting water was primarily a woman's chore and they had to do it at all times of the day; there are strong cultural connections between women and carrying water (Nasreen, 1995).

In the study area, the distance of the water source in the community such as tube wells, ponds and wells from households is between 500 yards and 3 km for most households. Because water sources in the community/neighbourhood are affected during a disaster, women are required to travel long distances. Research (Khatoon, 1995) shows that women will have to walk to collect drinking water for households because of the dwindling source of ground water and increase in the salinity of surface water. Women must walk longer distances to collect water, which is a challenging traditional restriction on women's freedom of movement.

Discussion with UDMC members shows that during a disaster male members of the family sometimes collect water because it is located very far from home. If women go to collect water from far away, it hinders other household-related activities such as food preparation. Respondents agree that women really encounter problems when they first collect water because of their dress *saree* (and *borkha* is even more difficult and is now used less), which are particularly problematic for walking long distances. Secondly, when GO and NGOs provide water, they usually stay on the riverside because it is convenient for them to come by boat. Women often walk a long way through contaminated water for hours to reach fresh water, leading to significant health impacts, including skin diseases and gynaecological problems. They train their children to drink less—a practice that has ongoing and serious health consequences reported in Alston (2015).

In addition to walking, a common method of transport to collect water is boat but women cannot row a boat because boating is outdoor work, and they require being physically strong to row a boat. To summarise, women are not used to boating because it is culturally unacceptable. This study shows that collecting water is culturally a job for women. Discussants were asked why women not men? One of the participants responded:

> Traditionally, women are responsible for household work. I saw my grandmother work and my wife saw her grandmother; this is the practice in the society. Almost within 1 km, there is a common deep tube well and pond managed by the community. Water is now available in nearby places because it is easier for women to collect water. (Male, 39, class XIV, FGD with UDMC)

However, cooking is only done by women; therefore, during a disaster, fuel wood for cooking became an acute problem because of the massive destruction of natural settings on which people in the coastal community are dependent. The vulnerability of rural women in developing countries who are often dependent on natural resources for their livelihoods, such as collecting water and fuel which are more likely to be the women's responsibility, has long been attributed to climate change (IPCC, 2007).

The *Sundarbans*, owned by GO, are the most common sources of fuel wood for villagers. The populations surrounding the *sundarbans* have high poverty rates and limited livelihood opportunities (O'Donnel & Wodon, 2015). Note that ~94% of respondents agree that fuel wood is a big crisis during a disaster. Male members do not take responsibility because they earn and work outside and are unwilling to take responsibility for collecting fuel wood for cultural reasons. Furthermore, because of fear of the forestry administration, forest security guards, and coastguards, males do not go to the forest and fuel wood scarcity turns into an extra burden for women. A woman describes the following:

> We are dependent on jungles (Sundarbans) for fuel wood. Sometimes, having no alternatives, women of the locality go into jungles for stealing wood and face many problems. There is fear of pirates, tigers, snakes, insects, and worms. Despite the fear of attacks by pirates and tigers, women go to the jungle for collecting fuel wood for cooking because cooking is the sole responsibility of women and males are not used to cooking. (Woman, 23, class XI, in-depth individual interview)

3.6 Disaster, Relief/Aid Distribution and Women

The practices of discrimination and injustices for distributing and allocating relief and aid are noted by respondents. This study shows certain issues such as nepotism, political affiliation, personal relations, conflicting views, competing interests, individual interests and family interests. These are actively considered at the time of allocating relief, aid and assistance, which creates discrimination and vulnerability, particularly for women; this is consistent with other studies (Fukuyama, 2001; Nakagawa & Shaw, 2005; Nasreen, 1995; Saha, 2015). The discrimination and unequal distribution of relief affects certain families and neighbours physically and psychologically during a disaster (Saha, 2015). In focus group discussion with UDMC, a male respondent asserts the following:

> We do not face any noticeable problem and gender discrimination in terms of providing relief. The relief card is in the name of women and women are in the priority list of different projects such as Kajer Binimoye Khaddo or Mati Kata of GO and NGO. How come you expect discrimination in distributing relief? There was no such discrimination on the basis of men and women; rather, women get preference. (Male, 52, class IX, FGD with UDMC)

However, from the survey data, 27% of respondents' state that they encountered gender discrimination in collecting relief, aid and assistance compared to the majority of the respondents (73%) who say they do not face any gender discrimination. However, subsequently, after conducting the FGD with vulnerable women, one of the woman respondents informs that relief goods are stolen by the UP chairman and members.

> UP chairman and members give us less in quantity and ask us to pay money. They prove importance to their personal relations. After Aila, everyone did not get equal materials for house development. They asked me to pay them 5,000 taka and I have paid it. After Aila, GO paid every family 20,000 taka cash but we got 15,000 taka only. We did not raise our voice and could not say anything for the fear of being excluded from the list of relief aid next time. There is discrimination in relief too. (Woman, 45, class V, FGD with vulnerable women)

When asked why you did not say this in the discussion? The woman respondent replied the following: *I did not disclose in the meeting as one of the UP members was close to the site of the focus group. I cannot say anything in his presence. They provide extra facilities to some people who have good relations with them and who belong to their political ideology. Some people who have no one in the family to raise voice are particularly deprived. Basically, people who are niriho are more discriminated against and getting less quantity in weight is extremely regular* (Woman, 45, class V, FGD with vulnerable women).

In this study, the corruption and nepotism of GO officials and UP members for distributing relief aid is noticed. Studies (Alston, 2015; Nasreen, 2008; Saha, 2015; Younus, 2017) reveal most *upazila* and union-based actions in relief and rehabilitation are primarily implemented by the UNO and UP chairman, and they create the list of recipients for aid. However, relief and aid are commonly allocated to people who are either political colleagues or relatives and to people who have positions on elected bodies.

Harassment of women in relief queues occurs after a disaster, which creates significant hardship for women. Respondents identify that discrimination does happen based on political identity (17%). It is matter whether the relief seeker belongs to the party/ideology of UP members and the party in state power (which is true in national politics). It does happen in different forms such as being kept waiting in the queue a long time (25%), being given less in quantity (16%), being asked for a bribe (11%) and nepotism (31%). Then, gender discrimination and relief distribution is described by a woman:

> I do not know much about gender discrimination to get relief and aid; however, I heard from other people that women are discriminated in getting relief and aid. For example, against a 10kg card, women are getting 8 kg. Local leaders and gram matabbor [a person from the community who is influential, aged, well-informed, conduct local justice, local elite and who has affiliation with leaders, administration, and political parties] those who are responsible for distribution do this sort injustice with women. (Woman, 23, in-depth individual interview)

The dependence on outside help in the immediate post-disaster phase leaves people highly vulnerable to food and water insecurity (Alston, 2015). There are multiple families and people who were solvent before disasters but became poor after losing everything to disaster. They were used to helping others and even had paid workers and servants. Men from these families do not like to collect relief because it is unpleasant and disgraceful for them. Study reports that because of their social status, the previously solvent family, particularly male members, do not want to get relief in front of others (Khondker, 1996; Vinas, 1998). However, respondents argue that, in desperate situations, women from solvent families can do it because they are considered inferior. The role of women from families who have lost a good livelihood has been described as follows:

> Yes, it is matter of shame and a question of dignity and inferiority for certain families to collect relief because they were well off before a disaster and lost everything overnight. Seeking help and queuing for aid is a prestige issue for them. However, GO-NGO gives priority to woman and provides additional privileges to women. Therefore, women are in the front and they must go for relief. (Male, 47, class XIV, in-depth individual interview)

This study shows that at the time of relief distribution and making a list for aid, the priority is the landless poor. However, *Aila* equally affected the rich and poor people and *Aila* created homelessness and landlessness across all people in the study area. Nevertheless, one of the participants shows the scenario of vulnerability of women by stating that

> NGOs give priority to the needy, destitute, and poorer people. Those who have land size of 2-4 bighas and other resources did not get aid and assistance from NGOs. However, the better off portion of the people had problems and led a worse life after Aila because they can not seek aid from NGO because of shyness. Note that Aila devastated everyone equally. (Male, 45, class X, FGD with vulnerable people)

3.7 *Purdha* System, Weather Warnings and Vulnerability of Women

Purdha is a curtain or veil that covers the body of the women, hiding their appearance and any part of the body from being seen by outsiders; however, its' social impacts are far greater than mere attire. *Purdha* is the socio-cultural structure and practices of society. In a strict sense, *purdha* is a way to restrict women to stay in households from the age of puberty and limit their mobility, which is defined as "not allowing women to be seen by strange men and to be covered in veils whenever they go out" (McCarthy, 1977 as cited in Nasreen, 1995). In a broader sense, *purdha* indicates women must be decent and modest and restrict their interaction with strange men (Feldman, 2001; Rozario, 1992).

Purdha maintenance in Bangladesh differs as per socio-economic category, religion, family background, level and type of education and age. The socio-cultural phenomena of Bangladesh assign roles, rights, and responsibilities between men and women and frame the relationships between them. However, *purdha* is at the heart of gender discrimination and division of labour. *Purdha* determines inside work for women and outside work for men in coastal areas (Chen, 1986). Women in Sutarkhali and Kali Nagor were asked whether or not they maintained *purdha*? All respondents report that they maintain *purdha*; it is the reality of society and it becomes part of their daily life. Even when people of the study area were in danger from cyclone and other natural hazards, they expect women to keep themselves decent and modestly dress. What *purdha* means for women in the study area and how it creates disproportionate vulnerability is reflected in the following statement:

> Basically, men wear small and light dress and women wear a saree, which is extremely common and traditional for women. Saree as a dress is extremely long and difficult to manage at the time of danger. Women cannot swim with a saree; moreover, they cannot quit the saree because of societal norms. Definitely, society and prestige is extremely important in the society than surviving. It makes women more vulnerable. (Male, 50, class X, FGD with PIC members)

Regarding the connection between dress and vulnerability of women, another male discussant reports the following :

> Men can struggle but women cannot do so. Men do not bother with societal norms; they can swim by being nude but women cannot do that as they care about norms. Men can run hard or leave their clothes behind, but women face much difficulty with their dress. Women are ready to die for maintaining their honour and society. (Male, 45, class XVI, FGD with PIC members)

A woman UP member describes:

> We faced multiple problems as women during a disaster because of socio-cultural and religious norms. Men can change their clothes in front of other men and can easily swim; however, we can not do so because social traditions are very strict for maintaining dignity of women. The seniors in the family forbid women to go outside. Eventually, women suffer while following social traditions. (Woman, 50, class XII, FGD with UDMC members)

Nevertheless, respondents of the coastal area report that the practice of *purdha* does not mean always cover the body of women with a *borkha*; rather, it means not appearing in front of male strangers at any time, covering their heads and going out with a veil, accompanied by male members and wearing dresses as per the dress code. When asked what are the problems of *purdha,* only 27% of respondents identify that *purdha* creates problems for women for their activities in response, particularly the activities related to the public sphere such a working outside, collecting relief and necessary goods, free movement problems, showering problems and not being allowed to swim.

Maintaining a *purdha* in a rigid manner is declining in coastal areas; however, the discussants report it will take time to change completely. Why will it take time? A respondent reports that the practice of *purdha* is similar to our culture. It has been practised for centuries and generation after generation. Kabeer (1990) reported that *purdha* is grounded in the structural constraints that deny women access not to both social power and any form of autonomy over their own lives and religious prescription (Fig. 3.8).

Weather forecasts and warnings are disseminated before any natural hazard event, such as a cyclone in the coastal area, using technology like television, radio, mobile phones and other sources of information. By staying at home and having social restrictions on their mobility, disaster-related warnings are delayed for women, which places them in riskier positions than men. Other research (Alston, 2015) reports that women are far less likely than men to receive early warnings because they may not have access to mobile phones, radios or televisions; moreover, they may be isolated in their homes and restricted in their movements by customary practices. In the study area, the lower rate of education, unemployment for women, male domination, the *purdha* system and restrictions on mobility for women limit the access of women to television and radio and other information sources. In the study area, very few households have television. In most cases, TVs are in the market, tea stall, club or *mor* where men gather for their leisure time, and women have restricted entry to such public places.

Nowadays, easier access and availability of mobile phones makes the dissemination of warning information easier and quicker in coastal areas. However, mobile phones belong to male family members and most time male family members are

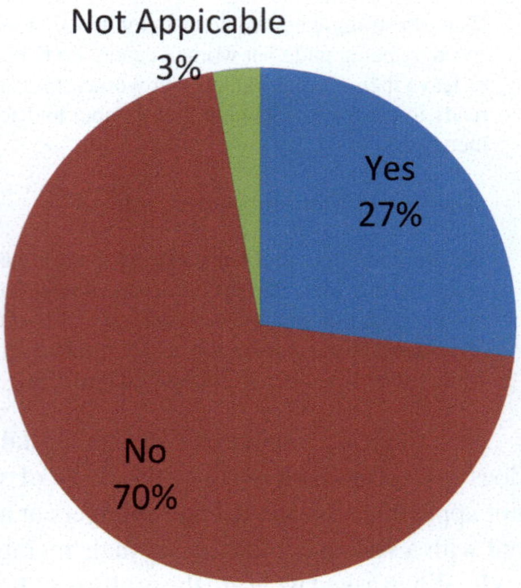

Fig. 3.8 Opinion of respondents regarding problems created because of *purdha*

outside the house for work or for business. A community woman answers is listed below:

> We have one mobile and my husband keeps it with him. Whenever he goes to work, in the bazar and in the mor for gossiping he carries the mobile with him. If someone calls for work or for some other purpose, my husband can answer better. I do not know everything. Indeed, people sometimes make calls by mistake (sometimes genuine mistake but later on they take a chance to disturb and talk with women and sometimes intentionally do random harassment) and try to disturb women over the phone. Certain husbands do not trust their wives; they believe that the wife may build an illegitimate relation with another person. (Woman, 45, Class XII, in-depth individual interview)

While examining the relationship between vulnerability and dissemination of disaster warnings (Table 3.2) in 80% of cases, it was men who received the warning earlier compared to only 5% of women. Patriarchal norms and *purdha* constrain and limit women from taking life-saving initiatives in a coastal community. Women are dependent on men to share early warning messages and have considerably more limited access to information. They may not receive warnings or men may not have the time or inclination to pass them on because customary law does not allow women to leave their compound without a male escort or at least without permission. Consequently, multiple women have perished because they do not receive information or do not have permission to act on the advice. The respondents report that the family with male members gets information and updated weather forecasts faster compared to the family without male members. A women respondent described the following:

> During the last durjug (Cyclone Aila), I did not know anything and I was sitting idle in my house. All of a sudden, I heard noises and sounds of rushing from neighbours, and then I came out. By that time, many of them started to go to look for safe shelter. I had no time to take anything; there were announcements from the mosque to go to the shelter centre as soon as possible. I left everything behind and went in the shelter centre with a single set

Table 3.2 Dissemination of hazard warning and vulnerability of women

Sources of hazard warning	N	%
Miking (instrument with loud sound)	98	37.3
E-media (radio/TV/mobile)	71	27.0
Announcement from Mosque	23	8.7
Male members	15	5.7
Neighbour	12	4.6
Bazar/tea stall	5	1.9
Others	5	1.9
Did not get warning	34	12.9
Receiving warning from various media by gender		
Male	209	79.5
Female	14	5.3
Did not receive	40	15.2
Reasons for receiving warning delayed by women		
Women staying at home	178	67.7
Warning in public place	44	16.7
Social restrictions	23	8.8
No access to e-media	18	6.8
Early warning to women reduce vulnerability		
Yes	251	95.4
No	9	3.4
Do not know	3	1.1
Total	**263**	**100**

Source Fieldwork, 2019

of clothes. If I had television or mobile and if I had a male family member in the family, I would receive updates from the mor and I could go to shelter with certain necessary goods. (Women, 50, illiterate, FGD with vulnerable female-headed households)

In Bangladesh, women who follow patriarchal norms and practices by keeping themselves within the house and maintaining the *purdha* system, are treated by the society to preserve the "sanctity of homes" and keep their husband's honour (Nasreen, 1995); furthermore, if they go out their honour may be impugned (Alston, 2015). In this regard, Kabeer (1990) argued that the most striking aspect in the organisation of gender relations in the Bangladeshi society is absolutely in the denial of autonomy to women in every aspect of their lives. The most important sources of information were *miking* (mobile microphones) (37%) in public places and throughout the roads where few women have access. Accessing radio, TV and other technology such as mobile phones was identified as the second largest source of dissemination of warning by 27% of respondents. In addition to these two sources, warning in *bazar*/tea stall (2%) and announcement from a mosque (9%) are identified as sources of warning where women normally had no access (Table 3.2).

Purdha makes women vulnerable not simply because they have less access to male-dominated shelters, but because it enforces women's dependence on men for advance warning and preparedness information. Although during cyclones, women are handicapped by the fear of shame attached to leaving the house and moving in public (Cannon, 2002; Ikeda, 1995); this has remarkably changed over the past few decades.

Almost all (251) respondents (95%) strongly agree that if women receive early warning information, the vulnerability to a cyclone will reduce. The reason why women are not receiving warning earlier than men are because of staying at home (68%), warnings being circulated in public places (5%), no access to e-media (9%), and social restrictions (9%) (Table 3.2). Furthermore, text messages having the latest updates are sent to all mobile phones from the centre by the GO. The dissemination of warnings in public places where women have less access because of social restrictions leads to disproportionate vulnerability of women. The following reflects the fragility and reality of information dissemination and vulnerability of women:

> Normally, warning is preached in public places. I received the warning 2h before Aila came, which was significantly less than that for others. However, there was a warning message in the morning on May 25, 2009 regarding the possible severity in the market and tea stall which is a gathering place of males. Because women remained at home, they received the news later in the day. Water started to flood the locality before the warning reached from the market and tea stall to homes; I did not have enough time to be prepared. (Woman, 43, can sign only, FGD with female-headed households)

The respondents report that access to information and information technology directly for women enriches the knowledge of women. The advantages of the information dissemination at the right time and the right person are reflected in the voice of a woman as follows:

> It is women who know everything about households and women know the exact location of the necessary goods. If we know early, we can carry necessary goods at the shelter centre and can preserve them after a disaster. The direct access to information and information technology help us realised the severity of a disaster and take measures. (Woman, 30, class XII, FGD with UDMC members)

Another woman describes access to information and advantages as follows:

> I cannot keep necessary things near my hand all time. If I know the update warning, I can keep necessary items nearby such as a burner such that I do not face any problem for cooking. (Woman, 33, class X, FGD with UDMC member)

This study reports that the mobility of women, education of women, training programmes and access to information and information technology help women taking sustainable and well-planned initiatives and adaptation strategies. For example, a respondent explains that, at Sutarkhali and Kali Nagor, many houses started to install rain-harvesting technology; most of this is maintained by women. Respondents report that this technology is cheaper and the best adaptive mechanism to overcome cooking and drinking water crises; this became possible because of training and information dissemination. Access to information and information technology increases the bargaining capacity and role in decision-making power in the family and locality. A woman UP member describes the following:

> As a UP member, I did not know how to perform my responsibilities. Male members and chairman did not help me as I expected. Once I knew that there is training on strengthening local capacity during a disaster and roles organised by an NGO, I attended meeting and received practical direct knowledge about the role and responsibility of a local leader as well as things to do in a disastrous time. I shared experiences with local women and they are benefitting. (Woman, 50, class XII, FGD with UDMC members)

> I heard multiple programmes are being telecasted regarding what to do in disastrous time; however, I cannot see them since I do not have access to TV. In our village, there is a TV in the bazar but only men can go there. If women can watch these programmes, their family will be benefitting more and it will increase the adaptive capacity of women. (Woman, 65, Class II, FGD with female-headed households)

To improve coping mechanisms and adaptation strategies, access to information and information technology for women is necessary. Most women did not have access to information in the study area; this constrains them from taking initiatives and consequently they cannot adapt to the climatic variability and climatic disasters in coastal areas. The involvement of women with the public at community level, public forums/meetings, committees, more involvement with development activities and training as adaptation strategies introduced by GO and NGOs are addressed in Chapter 6.

3.8 *Maya* to Household Belongings and Vulnerability of Women

In response to questions whether women have privileges to shift to shelter centres; whether women are being obstructed from taking life-saving initiatives is revealed by respondents and discussants that for *Aila* in Sutarkhali and Kali Nagor women are independent to make decisions to go to shelter centres by themselves and women were not obstructed from going to shelter centres. This study reports that 91% (240) of respondents report that they evacuated their residence and shifted to a shelter centre for the last cyclone *Aila*. In most cases, the respondents who did not shift to shelter centres had houses, which were sufficiently strong and high enough and other people took shelter in their houses. Furthermore, the distance of the shelter centre, backward transport systems and *maya* to households' goods did not allow them to shift into the shelter centre.

In addition to multiple reasons for attachment, many women are reluctant to leave their houses because evacuation may mean the loss of cooking utensils, goods and livestock; fear of losing houses and essentials; and lack of cooking and toilet facilities to ensure honour and family food security (Oxfam & Naripad, 2011 as cited in Alston, 2015). Ahmed et al. (2008) reported that when a hazard strikes women have to safeguard all belongings, send everybody to safer refuge and stay back to ensure that the household is not ransacked while the family is absent. In this process, women accept whatever consequences arise from the imminent hazardous event. In a quickly developing hazard event, an apparent "passive" response may lead to death; such a risk is deliberately undertaken by women. The respondents report that sometimes women went into danger because of their attachment to household goods, livestock, and poultry. A woman respondent described sometimes women members of households did not go to shelter centre since:

> Women have different feelings about all belongings compared to men. Women feel like crying if they have to leave their households; for example, the cows and poultry of the house she looks after. During Aila, my son's wife did not leave the household because she did not want their wealth to be left unsecure. To summarize, maya grew in women. (Woman, 45, class V, FGD with PIC members)

A male UDMC member describes the following:

> Many women do not want to leave houses and wealth unprotected; they have an extremely strong attachment to their wealth. Women take care of ducks, hens, and cattle all the time. They jeopardise themselves to protect domestic animals. Men only earn but women do compulsory tasks for preserving and maintaining family; this is tradition. Women are used to staying at home. (Male, 45, class XVI, FGD with UDMC)

Respondents report that with certain exceptions at present women normally are not obstructed by male members and social norms. Even women of this area are not dependent any more on the decision of male members and males do not want women to stay at home and wait for their decisions in crisis, particularly after *Aila* in 2009. What leads to changes in coastal areas? Nowadays, every family has good communication systems because of the availability of mobile phones, and it is easier to pass information received by mobile phone and women do not wait for male members to come in person to rescue women and take women in safe shelter. The waves of change in coastal areas are now manifested and became possible because of the rapid expansion/availability of technology (such as mobile phones, radio, TV and community radio), in addition to the large-scale intervention of GO-NGOs and empowerment projects for women.

This study reports that only 11 respondents out of 263 were obstructed by family members/neighbours (6), religious leaders (1), community leaders (2) and others to go into the shelter over the past ten years, particularly after *Aila*, thus reflecting changes in social practices. The focus group discussion (FGD) shows that in terms of the absence of a male person, woman went out for safe shelter and did not wait for a male's decision. Accordingly, a male member commented about the changing attitudes of male members towards women:

> During last durjog (cyclone Aila) I was not even at home. If I were at home I would definitely first take women members to the safe centre. Because I was not at home, my wife managed everything and took children and other emergency goods with her to the shelter centre; she made all these decisions. (Male, 32 year, class X, FGD with PIC members)

In the FGD with UDMC, all respondents confirmed that people in their community are much more conscious than ever before. For example, people used to ignore warnings and signals and were not in the habit of going to the shelter centre. However, people are willingly going to the shelter centre and people have recently become conscious and heed the warnings about cyclone in the locality.

3.9 Female-Headed Households and Vulnerability

In the study area, women normally become the head of the household when they are divorced, separated or widowed; there is no mature adult male member in the family or if women have only disabled and have sick male members in the family. In the study area, only 5% of households are nationally headed by women compared to 16%. Many female heads of households who have been divorced or widowed are culturally discouraged from remarrying as per the respondents, which is consistent with Cannon (2002).

Research (Oxfam, 2011) shows that disasters reinforce, perpetuate and increase gender inequality and make bad situations worse for female-headed households, which has been unveiled in the study area too. The problems for a female-headed household seen in the study area are multiple, including lower wages compared to men, discrimination in getting GO aid and relief, illegal offers by neighbouring males, fear of sexual harassment, discriminatory attitudes by society, lack of guardianship, lack of ties with husband's family and denial of resources by the husband's family. These factors pushed female-headed households into a discriminatory situation. In response to the question "do female-headed households face more problems compared to male-headed households in disastrous time?" almost all (92%) respondents said that female-headed households face more problems and they are more vulnerable compared to male-headed households in the study area. The respondents' report that the discriminatory attitudes and vulnerability of female-headed households are manifested more during a crisis.

They argue discriminatory attitudes to female-headed households becoming visible in daily life activities starting from individual level and continuing to family and community level. For example, if a woman wants to sell her labour to earn money, she is given a lower wage; moreover, a working woman in a rural set-up is often sexually harassed because it is perceived that a woman only accepts a labourer's job when she does not have a male relation and protector to feed her. Therefore, others take advantage of her helpless condition and certain people want to take on a pseudo guardianship of women with an ill motive. These factors create a vulnerable condition for women who are widowed and must work and will do whatever is available because there are limited or no options (Ahmed et al., 2008; Alston, 2015). One male respondent states the following:

> Female-headed households, widows, and abandoned women [women who are not cared by husbands and a woman who does not know where her husband is] are extremely vulnerable. Sometimes, they have been feeling abashed by the behaviour and talk of influential persons of the community and certain bad people who took advantage of an opportunity. Widows and female-headed households get problems going to the cyclone centre even if they manage

> to go they do not get any place and people do not care for them. (Male, 49, class X, FGD with PIC members)

All respondents of the focus group with vulnerable villagers and vulnerable women, irrespective of being man or woman, agree that a female-headed family is always seen as helpless in their society; sometimes, even their family members do not welcome them. Vulnerable women of female-headed households report that they are sometimes treated as if they have no guardian because there is no male member in the family. Because of helplessness and a lack of feeling protected, female-headed households possibly have lower income and a weaker asset base. Women from female-headed households are disproportionately discriminated against when they attempt to access resources, credit, employment opportunities, education and social security, which is acknowledged by survey respondents and focus group discussants. Female-headed households in which there is no maturing male or male earning member have more trouble in surviving.

Respondents report that the quality of housing, location of housing on raised ground/high land and adequate food storage are important elements during a disaster for self-protection, but female-headed households have more difficulty achieving them without any male help. Women's ability to create safe conditions in the face of impending floods or cyclones is reduced. During a disaster, everyone in the study area keeps themselves busy with their own studies, and conditions and atmosphere place women of female-headed households in the "vulnerable within the vulnerable" (Ariyabandu, 2010) category. In terms of the vulnerability of female-headed households, a woman reports the following:

> During a disaster, families who have no male members, particularly female headed households, suffer more than those who have male members in the family. Women cannot rush and manage everything during a disaster. If the family has got children and aged people, they are then extremely vulnerable and become a double burden for women. It becomes extremely difficult for a woman to manage all these aspects. When all community members are in trouble, it worsens for female-headed households and it is difficult to get help from others. (Woman, 25, Class VI, FGD with CPP members)

The spaces women occupy are socially delimited; in particular, women (as mothers) are traditionally excluded from or have limited access to the domain of power: the public space of waged labour as is unfolded by FGD respondents. Female-headed households often lack support from social networks and state (Fordham & Ketteridge, 1998). In response to the question "how the female-headed households are more vulnerable compared to male-headed households as a disaster unfolds?" respondents in the focus group with female-headed households show that the workload increases but income from the female-household member gets reduced.

In a female-headed family, women are not only accountable to solve household-related responsibilities and look after dependent family members with the least access to resources and assistance, they have to earn. Note that these women have to compete with men for incomes in a situation when employment opportunities are reduced because of a disaster. Wiest (as cited in Enarson, 2000) and Saha (2015) report the same problems for single mothers in Bangladesh. In terms of discrimination in wages, the respondents in the focus group report the following:

> There is wage discrimination for women in the private sector, e.g., in shrimp gher. Women and women from female-headed households who work in shrimp gher are paid a smaller wage compared to males. For males 300 taka is paid for per day compared to 200 taka for a woman in a day. Moreover, the payments for women are paid in multiple instalments. (Woman, 25, class XIV, FGD with vulnerable women)

Is there any discrimination against female-headed households when they attempt to get support from GO? In response to this question, respondents of the study area argue that female-headed households face certain additional specific problems than a family with male members. Furthermore, 29% of respondents report that female-headed households are discriminated against in getting relief while 14% identify illegal offers from neighbouring males and 39% claim that female-headed households get less in wages, whereas only 4% and 3% identify ignorance by GO and NGO, respectively. In terms of the greater vulnerability of female-headed households, NGO activists report that for a GO aid card during and after a disaster, women have to go to the UP chairman and members. Furthermore, the chairman and members have asked for bribes but no one confesses this publicly; however one of the respondents reports the following:

> Yes indeed there is discrimination and abuse of aged women and widows; however, these incidents happen in secret. Different people in politics have taken the opportunity of weakness. Certain women may have to get engaged with physical relations, particularly those who are widows and destitute women. They lure destitute women by offering a relief card or elderly allowance card. (Male, 48, class XIV, FGD with NGO activists)

An aged woman respondent describes her vulnerability as follows:

> I am extremely vulnerable because I am alone and have no husband. Nobody care about me because I cannot raise my voice and I have no one to even raise voice for me. Look at the house and location of the house that I got from the GO. My house is the worst and the surrounding is flooded when it rains. Everyone selected houses for themselves and I got the last one. Everyone is cultivating vegetables around their house; however, I cannot do gardening because of water logging. (Woman, 59, illiterate, FGD with female-headed women)

Despite challenges, including lower wages, religious ignorance, harassment and exploitation by local influentials and being deprived of aid and assistance, women from female-headed households are involved with earning activities to manage and meet the requirements of their households. In addition to earning-related activities, the head of female-headed households looks after their family and takes care of dependent family members. Despite socio-cultural and religious obstructions, women of female-headed households are used to working outside and are involved with income-generating activities; fishing, day labouring, *mati kata and shaola bacha* in the shrimp *gher* are the common activities. Participants from this study report the coping mechanisms of women from female-headed households, e.g., they start working in the *gher* early morning and late at night and they finish their work before sunrise such that no male members know women are working in the *gher* and women do not need to work with male members.

This study reports that women who have young children have limited access to waged labour; moreover, these mothers are dependent on aid and assistances from relatives (24%), allowances such as the widows' allowance, old age allowance from GO and NGO (8%), doing home-based income-generating activities such as sewing, vegetable gardening, poultry and livestock after being trained by NGOs (14%). In terms of working outside as adaptation mechanisms, a woman respondent argues the following:

> Life is first. If I do not work who will give me money to take care of my family. People who are talking about my activities are they helping me? If I earn money in a proper way to survive and if I work honestly, following religious principles nothing is wrong. (Woman, 37, class V, FGD with female-headed households)

Chant (2008) reported that female heads may lack ties with their ex-partner's relatives and with their own families and communities, which is reflected in this study. Respondents report that when a woman does not identify no ties, no respect, or no dignity, she does not feel honoured anymore to stay in her husband's house and therefore migrates to her parent's house as her last opportunity for shelter, which is reported in this study by 8% respondents. As a coping mechanism, why do women of female-headed households migrate to their parent's house?

> It is not extremely common but sometimes family members do not want a widow staying in her husband's house. A dead husband's family members think of a widow as a burden. Moreover, if she stayed with them in future, she will get a share of the property. Furthermore, bad people tease and disturb women who do not have husbands and sexually harass them. In that case, women from female-headed households take shelter at their parent's house. (Woman, 45, class XII, in-depth individual interview)

Nevertheless, the undermining situations of women all over the country, including the study area, are improving. From 1998 to date, different social safety programmes such as husband-deserted women allowance (BDT500 for a person per month), widow allowance (BDT500 for a person per month), old age (62 years and up) allowance (BDT 500 for a person per month) and destitute women allowance (BDT 500 per person per month) continues by the Bangladesh GO under the Department of Social Service and to date at large scale, respectively, 6,900 million BDT and 1,890 million BDT in 2020 has been given (www.dss.gov.bd).

3.10 Summary of the Chapter and PPVM Model

This chapter briefly provides a fore taste vulnerability of coastal women in the context of gender and gender relations in their daily lives. This shows that the coastal society is a highly gender-segmented society because of social customs and practices where women are discriminated against; moreover, they are controlled and restricted in their everyday work starting from the household. The study results reveal that the existing socio-cultural practices leave women in a more vulnerable condition during a disaster in the Bangladeshi coastal area. Furthermore, there is greater marginalisation of women in the context of physical and mental health issues, access to information, greater restriction on following religious practices, more obligations to collect drinking water and fuel wood, a tendency to eat less and last and a higher malnutrition rate.

Women are restricted, discriminated against and controlled in the coastal area; however, women have a greater role in caring for the sick, injured children and the elderly, helping male household/family members in outside work, in addition to household activities that create a double burden on the shoulders of women. Disproportionate vulnerabilities for women are demonstrated in the coastal area because of socio-cultural and religious restrictions, particularly for young women and girls. Many women are reported to be distressed not only because of poverty and scarcity of food, lack of employment opportunities and financial hardship but also because women in the coastal area are distressed by traditional practices followed by society and family, as well as for migration of male members and absence of male members. The progression and pathways of vulnerability of coastal women in Bangladesh can be described in Fig. 3.9.

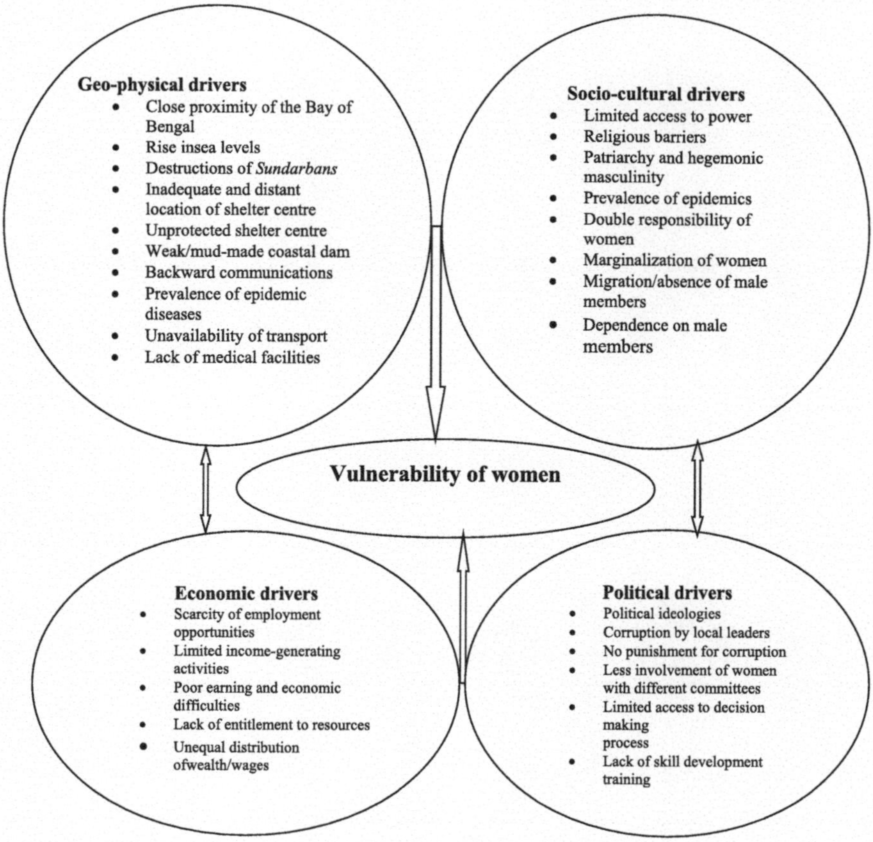

Fig. 3.9 Progression and pathways of vulnerability model (PPVM)

Chapter 4
Economy, Entitlements, Coping and Adaptation Mechanisms

4.1 Introduction

In male-dominated and traditional Bangladeshi coastal area, most material and non-material assets are owned by men; women cannot always have access to their husband's wages (Dreze & Sen, 1989; Jinnah, 2013; Kabeer, 1990). Women-related initiatives to exercise control over material and non-material assets are necessary for their empowerment (Panday, 2016). This study reports that women in the coastal area are deprived of their entitlements/access to resources and ownership of land; moreover, they can face discriminatory wages too. Consequently, women cannot play a role in decision-making either at a family or at a community level (Briceno, 2010). Successful coping/adaptation is not possible if people feel unsupported or at risk and if institutional support is not forthcoming (Tscharkert et al., 2011 as cited in Alston, 2015). Furthermore, gender inequalities for employment and income, ignorance of entitlements and access by women to resources are perpetuating social degradation and reducing capacity (Enarson, 2012) in coastal Bangladesh, although women's "double shift" work is central to adaptation and sustainable practices (Alston, 2015). This chapter focuses on both vulnerability and adaptation mechanisms adopted by women in coastal areas, especially in terms of lack of entitlements and lack of access to resources. The manner in which male migration and unemployment in the locality introduces vulnerability of women, including sexual harassment and insecurity issues for women in the coastal area, has been addressed in this chapter.

The respondents report that the challenges and problems in disastrous times are very similar throughout the study area but the responses to disaster and vulnerability differ. Therefore, the disaster-related impacts were not equal for all people in coastal areas. However, discrimination against women not only accentuates women's vulnerabilities during disasters, it also wastes women's potential as sources of resilience and adaptive capacity. In the study area, gender is an important factor for determining access to resources, decision-making power at a family level, community level, local level, involvement with different committees and access to different aid/assistance allocation and distribution processes. The respondents report that unequal access

M. M. Khan, *Disaster and Gender in Coastal Bangladesh*,
https://doi.org/10.1007/978-981-19-3284-7_4

to resources, lack of involvement with decision-making processes, lack of access to aid and assistance allocation and distribution programmes impacted the coping mechanisms and adaptive capacity of women in coastal communities.

Some respondents report that women feel excluded from the family and from society; this feeling of exclusion leads to the deterioration of confidence and capacity of women. Even within the same household, responses vary during a disaster because of gender-based roles and responsibilities. In FGDs, both women and men state that, in addition to socio-cultural factors, responses to challenges vary in the family because of the different physique and strength of men and women.

4.2 Disaster and Involvement with Earning Activities and Women

In this study, access and entitlements to resources include the access of women to micro-credit, loans from NGOs, loans from *samity* and *sangstha*, savings by forming a *samity*, any income received by the women, right to properties, ownership of livestock and poultry and income from them and ownership of small businesses. Despite extraordinary advances in women's status compared to the last decade, such as access to education, improvement of health and involvement with decision-making processes from national to local level, women in Bangladesh own very little of their country's land despite their extremely active participation in agricultural production and labour (Alston, 2015).

The vulnerability of women in coastal areas can be reduced by involving women in income-generating activities and by ensuring their access to income, which has been reflected in other studies (Alston, 2015; Saha, 2015; Zaman, 2009). The respondents report that the status of women will be improved by diversifying their sources of income and by building assets held in their names. How lack of entitlements to earning creates problems within households has been described by a woman in the following manner:

> Definitely my husband will take my earnings; however, I think whatever the amount of money women are earning, they should own at least 50% of that. If women own money, they can spend it for purchasing clothes for kids and for treatment; women can after all spend on needs. However, in most cases, women cannot spend money on requirements since the husband owns all the money and husbands spend the money as they wish. A husband does not look after all the requirements of his wife. For example, if women do not own money, they cannot buy good clothes, which are humiliating for women, particularly when she participates in social programmes. Old clothes keep a woman belittled by her relatives. However, the life of a woman is sometimes risked because women's medical treatment is ignored and women cannot receive their treatment if their husband does not want it. (Woman, 23, class XI, in-depth individual interview)

This study reports that 70% of all respondents have involvement with income generation. The respondents who were not involved with income-generating activities are from comparatively well-off families; they do not require to work for money or are aged or too young to be involved with income-generating activities. In FGDs with vulnerable women, discussants report that they have income and they earn money by working outside, by doing *samity*, vegetable gardening, rearing poultry and farming livestock on a small scale; however, women mostly gave money to their husband. The involvement of women with income-generating activities and their contribution in the study area is reflected in the following statement of a woman respondent:

> I got involved with income-generating activities because my husband alone cannot carry all expenses of the family. I gave my money to my husband and later on if I needed money from my husband, I got it. (Woman, 33, class VIII, FGD with vulnerable women)

In addition to household-based earning activities, women in coastal study areas were involved with labouring jobs, particularly with jobs that are mostly manual labour and do not require any special skills/training. There are certain jobs such as rowing a boat or trawler for communication and fishing in the river/sea and operating a tractor for cultivating land. Labouring jobs include fishing in river or constructing isles and boundaries of *gher.* Shrimp *gher* indicates farming/cultivating shrimp either in a small pond (locally called *tob*) or a large lake. These are mostly privately owned by an individual or family. Shrimp farming in the coastal southern part of Bangladesh has rapidly and massively grown because of favourable climatic conditions and adequate saline water. Furthermore, women participate in *mati kata,* digging canals, constructing dams or doing small business in local areas. However, cyclone *Aila* shattered/limited opportunities that are explained by a male as follows:

> Men can continue earning money by migrating from their locality but women cannot. There are no agriculture-related activities and shrimp gher even without fish in the river for fishing. Cyclone Aila inundated agricultural land with saline water and washed the shrimp gher away. Therefore, women were in financial hardship. (Male, 49, class X, FGD with PIC members)

Similarly, a woman describes migration-related problems with their normal work opportunities:

> Women have no freedom to move and migrate somewhere else; however, men can move wherever they would like. Women are confined at home doing caring and household related activities for family. Sometimes, women earn money from home-based earning activities but husbands take that money too. (Woman, 30 years, can sign, FGD with vulnerable women)

Comments made in FGDs regarding access to earned money and the vulnerability of women were supported by survey data that reflects 11% of women independently spend money compared to 64% of men, whereas 25% of both men and women spend money together. The respondents report that, in the study area, women to some extent have access only in requirement and with their husband's consent. Spending money without the agreement of their husband is in violation of norms reported by respondents. One of the participants reports the following:

> I earn money by rearing poultry; however, I am accountable to my husband while spending that money. (Woman, 22, class IX, FGD with vulnerable women)

To overcome financial challenges, women get support from neighbours and relatives (23%), eating less (24%), and spending money from saving (13%). When asked about savings, participants responded that women always save certain money in the study area, which they never disclose to husband and male family members. In most cases, women save money in an earthen bank and maintain that hidden and spend it during moments of crisis. Studies (O'Donnell & Wodon, 2015) report that in the absence of saving, households borrow from friends and relatives or from financial institutions, which is consistent with the study results.

In terms of barriers to spending money by women respondents, factors such as patriarchal system (24%), lack of rights of women (30%), domination of the husband (13%), males being main family earners (8%), less income by women (8%), males knowing better (8%) and social norms (6%), are identified. In terms of barriers a woman describes:

> It is the society's rule that women will have fewer rights to their earnings compared to male members because it is a male-dominated society. People in coastal areas are extremely pious and fear religious norms. From religious and social perspectives, earnings must belong to the head of the household, and the majority of the heads of the household are male members. (Woman, 45, class XII, in-depth individual)

During a disaster, women's responsibility for their households and families continues with the expectation that they will protect property, shepherd children and livestock to safety and examine the family's survival, including building makeshift shelters and planning food security, which is noted in other studies such as Masood & Wahra (1997) as cited in Alston (2015). However, because of shattered income sources after *Aila*, it became a greater burden for women to meet household requirements. Their role and responsibility increased to some extent and the role and responsibility changed too. The changed responsibilities and greater burden are described as follows:

> Previously, because of the durjog (cyclone Aila), we lost everything, including houses. I was used to performing only household activities before; however, now I am busy with business (shopkeeping) for the whole day. I leave my children at home. Neighbours look after them, but I feel yet always tension about them; although sometimes I bring them to my shop. (Woman 37, class V, FGD with vulnerable people)

In this study, the high majority of respondents (84%) report that after *Aila*, they became involved with income-generating activities to overcome the family's financial crisis. The respondents report that almost every woman did something to earn money because there was an acute crisis and men's income was insufficient to cover family costs. While certain women did not get involved with any income-generating activities because of cultural or religious factors, they were from a wealthier family and work would reduce their prestige in most cases. One woman respondent reported the following:

> It is disgraceful for a woman to work outside and go outside the family because involvement with work outside hindersthe dignity of arich family. Moroever, it is unpleasant for certain families to let women members work outside because the leader of the working team is in most cases male. (Woman, 25, class 14, FGD with vulnerable women)

Khondker (1996) reported that disasters such as floods increased women's domestic burden. Moreover, disaster increases women's domestic burden by undermining women's well-being because of their dependence on economic activities linked to the home. The loss of utensils and other household essentials is a great hardship; moreover, floods undermine women's well-being in general. This was the case after cyclone *Aila*, which increased women's burden and created hardship because of the loss of household essentials, livestock, poultry and homestead gardening. Women suffer from additional difficulties because these all are common sources of income for women. Women who are dependent on poultry and livestock for income were more vulnerable in Sutarkhali and Kali Nagor after cyclone *Aila*. Family well-being was threatened when resources on which women relied to perform their important roles and obtain supplementary incomes was undermined. Poor women disproportionately suffered from a lack of food, clothing and shelter.

To summarise, the study results show that the amount of income of women is associated with their vulnerability and that is inclined to shape the coping mechanisms and adaptive capacity of family and individuals. In the study area during a disaster, poor women could not take better measures for adaptation, and poverty discouraged women from taking actions. Bolin and Bolton (as cited in Enarson and Morrow, 1998) report that women, especially low-income women, often poorly fare during the reconstruction phase. The poor, of which women are the majority, have less insurance, less savings and thus a less likelihood of a full, long-term recovery. The disaster-related impacts in the post-disaster period are more difficult to overcome, particularly for women.

> Yes, poor people suffer more compared to a wealthy family during a disaster and I believe it is true everywhere. The number of poor people in this community is quite high. A majority of people belong to the hand to mouth category; they do not have any savings to depend on. They take loans to meet requirements such as food in the off season when working opportunity is extremely limited, e.g., during the rainy season. (Woman, 39, class IX, in-depth individual interview)

Alston (2015) reported that women and girls are important agents of change in circumstances where climate change is creating major challenges and is important to the global battle to preserve our planet. In Sutarkhali and Kali Nagor, women are agents of change too and they are taking initiatives to deal financial crisis. For this purpose, along with home-based earning activities, including poultry, livestock rearing, sewing and vegetable gardening, women in the study area are becoming involved in outdoor works for earning such as *mati kata*, fishing, *deribadha*, *shaola bacha* and doing *jungle activities* (collecting honey, fuel wood and crabbing) to earn money. These tasks are market-related. Neelormi (2010 as cited in Alston 2015) reports that women express a strong desire for income-generating work, which is work that they can do close to home while attending their various activities. In this regard, a male respondent reports the following:

> Although it is better to have men in matir kaj, GO and NGOs give preference to women. It was the practice of this area that women stayed at home; however, at present women have started to come outside to work and they are earning to face hard times. (Male, 34, class X, FGD with CPP members)

Another woman explains the following:

> If I did not get mati khater kaj, I must borrow and take loans from other people but I am not doing that. The amount of money I am getting in every two weeks is enough to buy necessary goods and I am not seeking help from others. (Woman, 39, class VII, in-depth individual interview)

Women in the study area report that involvement with earning activities and earning at any cost either working at home or outside home, the availability of cash in hand, helps in overcoming financial crisis and this financial crisis determines the nature of the adaptation mechanisms of women. One woman NGO activist explained the following:

> I have a job in an NGO and have a fixed monthly salary. With my income, I bear some of the expenditure, including our children's education. I do not wait for my husband to pay my children's fees any longer. I can solve problems myself; all this became possible because my husband gave me freedom and trust for financial issues. (Woman, 35-year, class XIV, in-depth individual interview)

To overcome hardship, women got involved with earning activities both home-based and outside the home, including day labouring (56%), home-based income-generating activities (35%), fishing (20%), as well as migrating to other areas to earn money (1%), although the number is extremely small. This revealed that women in the study area started even taking a loan (15%) from NGOs and informal sources such as *samity* to start small businesses as well as for small poultry and livestock businesses that economically empower women. The economic empowerment of women improves their ability and supports the betterment and well-being of the community as argued by respondents. In terms of the involvement of women with earning as a coping mechanism, a male participant reported the following:

> Before Aila, women of this locality did not work outside; however, after Aila they have started to do so. Women are now getting engaged with activities of different GO and NGOs. Earlier, women used to maintain the purdha system but they faced food scarcity of and other hardship after Aila. To overcome scarcity of necessary goods and financial hardship and to pay for loans from sangstha, samity and dadon, women go outside and get involved with earning activities. (Male, 55, class VI, FGD with PIC members)

The changes in attitudes are reflected in the pattern of coping mechanisms of women to overcome financial crisis. The images show that women of Sutarkhali and Kali Nagor village are working as day labourers, including *mati kata* and agriculture work (Fig. 4.1), which are outdoor works. The outdoor works such as in shrimp *gher,* fishing opportunities or *mati kata* were devastated because of cyclone *Aila* but started to return to partial normality in two years.

Many women in southern coastal Bangladesh work outside the home and breaking down old customs, practices, *purdha* and male control because of circumstances, particularly from financial crisis. They are doing this work with the consent of the male guardians of their family, including their husbands and without male support; moreover, they are not likely to do so. Men allow women to work outside because the women's incomes are spent for family support and invested in income-generating

Fig. 4.1 Involvement of women with different earning activities

businesses. Moreover, men see the increase of income as beneficial for the family. Income from women working outside changes the expectations of gender relations, gender roles and social relations as well as in the study area that created a favourable atmosphere for changes in long-standing gender relations. In terms of changes in gender relations in the study area, a male respondent reports the following:

> Before Aila, I had cultivable land and I was able to meet the expenses of family members by cultivating land. However, after the tropical cyclone Aila, I became poor overnight and lost everything, including my land, crops, and livestock. Then, I started working as a day labourer, which was insufficient to meet up expenses of my family. The situations made me allow my wife to go for relief and work for money because we required surviving. (Male, 36, class VII, FGD with vulnerable community people)

Alston (2015) reported that there are instances where old meets novel and where non-traditional work meets customary law. Often, necessity overcomes customs and women must work to survive; this is a process for women to make adaptations.

The majority of survey respondents and focus group discussants of this study report that there are changes in local practices, values and attitudes, particularly towards women and their confinement in the house. Males in the community became flexible regarding the mobility and involvement of women with earning activities.

4.3 Disaster, Male's Migration and Women

This study suggests that crop failures because of flooding, salinity and tidal surge along with cyclones lead to more migration because coastal residents depend on the local resource base, which was just massively destroyed along with their livelihood. Survey participants and FGD respondents report that one of the advantages for men who cope with the disasters in the study area is to change and diversify their income sources. Moreover, in the study area, the most common strategy is to migrate other local places or cities. However, women cannot migrate even if they want to because of their gender identity and explain the respondents. Because women cannot migrate such as men, they face inequalities in accessing technology and power relations in both household and community, and dealing with gendered policies and institutions, which determine the levels of vulnerability. Leighton et al. (2011), Alston (2015), Mueller (2015) (cited in Nguyen & Wodon, 2015) and Saha (2015) report that many may be displaced or increased their reliance on migration as a coping strategy for survival and source of income to purchase basic necessities and send money back to their village.

Disaster acts as a "push factor" in the decision to migrate, and it can be attributed to destruction of livelihood options and houses, landlessness, poverty, lack of optimism for improvement and insecurity of their children, as reported by respondents and reflected in other studies (Bates, 2002; Kartiki, 2011) (cited in Alston, 2015). Improved employment (seasonal jobs and non-seasonal jobs) at the destination and lack of employment opportunities in the place of origin are the primary reasons for migration (Nguyen & Wodon, 2015; O'Donnell & Wodon, 2015).

Note that 89% of respondents report that disaster forces male members to migrate to another area to earn money; furthermore, the migration destination of male members ranges from Dhaka (capital city) to the nearby city of Khulna with extremely few moving to India. When examining the strength of cause and effect in the relationship between disaster and migration, cyclone *Aila* emerges as the central predictor of migration in the study area. Furthermore, in the wake of a disaster, migration is fundamentally influenced by the socio-cultural context in which people are embedded. Respondents report that male members primarily are pushed from the community, and this migration is both temporary and permanent depending on the effect level of the disaster.

The households were more affected by cyclones and who could not manage livelihood costs working in the community/local area after *Aila* have higher levels of migration compared to less affected households argue the discussants. This indicated that fewer women migrate from the coastal area compared to men because of concerns of social safety/security, the nature/uncertainty of jobs in the migrated area and the caring role and responsibility of women to family members left behind by males.

In response to why male members migrate, it is revealed that women cannot settle in any environment such as men. A man can stay even in an open place but a woman cannot stay in an open place such as a bus terminal, railway station or in a *bazar* because she requires social security and a separate arrangement to minimise problems such as sexual harassment, violence and rape. Furthermore, women cannot perform rickshaw and van driving in an unknown place with unknown people; even if they work, they might receive low wages or no wages. This study shows similar to other studies (Bryan et al., 2011; O'Donnell & Wodon, 2015) that women as vulnerable groups are unable to migrate because of uncertainty and fear of unsuccessful migration because of lack of safe and secured arrangements by society, in addition to poverty. This study reveals that unplanned migration and migration without prior arrangement by women maximise risk. The reasons for less migration of women is described by a man and a woman as follows:

> Women cannot migrate to other districts; it is difficult for women to work in other districts. Who will provide them security? Women do not migrate for lack of security. (Male, 47, class XIV, in-depth individual interview)

> Extremely few women migrate for earning in other districts from the coastal area because their husband and family do not allow them. For emergency and obligatory incidents, women migrate in other districts accompanying husbands and other male family members because social customs do not accept free mobility of the women. (Woman, 29, class IX, in-depth individual interview)

However, in addition to the mechanism for coping with financial crisis, women are involved with income-generating activities, particularly after *Aila,* and some of them have even migrated (1%) although the numbers are low. Survey results (60% of respondents of this study) and other research (Alston, 2015; Harvard Humanitarian Initiative, 2010) reports that women do not migrate similar to males in the coastal area because of the nature of their responsibilities and socio-cultural factors, which make it difficult for women to independently migrate, particularly young women.

The discussants report that woman members who migrated with male members from Sutarkhali and Kali Nagor were those who had nothing left after cyclone *Aila.* They were basically from the landless marginalised group who had small cottages that were washed away. They were dependent on daily earnings and because cyclone *Aila* devastated all livelihood options in the community these people moved somewhere else. However, people who had cultivable land, shrimp *gher*, their own house, business or other involvement did not migrate from the locality.

Furthermore, 38% of respondents say that they did not migrate because of patriarchal society and they identify in the FGD that sometimes even if husbands agree, the father-in-law and other male family members do not allow them to migrate. Moreover, religious barriers (42%) restrict women from migrating for earning; however, 20% of respondents interestingly reported that women did not migrate only because they are treated as weak physically compared to men. Certain respondents report that the dependent care responsibilities of women make them less mobile compared to males and are less able to migrate outside the affected area. Schmuck (as cited in Cannon 2002) argued that women's mobility is restricted because of their responsibility for children.

In this study, 66% of respondents report that the migration of males after a disaster creates extra responsibility for women and that extra responsibility includes caring for family (67%), economic responsibility (51%) and working outside the home (29%) for extra earning to meet their expenses. This study reports that mismanagement, unorganised and unplanned migration and displacement of males cause vulnerability, particularly for women in Sutarkhali and Kali Nagor village. In terms of the migration of males, this study reported that the migration of males from households places extra responsibility on women's shoulders. Out migration leads to significant tensions as men are extremely aware that this may leave families vulnerable and their wives without a male escort (Alston, 2015). The vulnerability of women in the absence of males is described by a woman respondent as follows:

> Migration of male family members causes considerable suffering to women. My husband has gone to India for work. In his absence, I have to face considerable pressure in managing family with extremely limited income of mine. Very often I cannot fulfil the demands of children. (Woman, 32, Class VIII, FGD with vulnerable women)

Another male respondent reported the following:

> In the absence of males, it becomes difficult for women to maintain their family. The presence of the husband leads to considerable differences. The absence of a male leads to many problems for women. For example, male members who migrated sometimes cannot send money for their wife and children and do not get news of their children and family. Whentaking care of children along with household activities, a burden for women is created. (Male, 47, class XIV, in-depth individual interview)

A woman reports:

> When men migrate to other districts, their wives have no scope to ask for money; women rather wait for money from husbands and money is sent sometimes irregularly and sometimes it has stopped for ever. It then creates a burden for women. (Woman, 25, class XIV, FGD with vulnerable women)

Respondents in the focus group report that frequent disasters in the coastal area causing wide spread migration of males result in women suffering from different risks, including sexual violence and harassment against women. The most common problem for women in coastal areas after male migration is identified as social insecurity (30%); this indicates that women feel vulnerable and unprotected in the absence of males in the household, whereas 15% of respondents report that they are sexually harassed by outsiders and neighbours. In the absence of males, neighbours sometimes come forward on different occasions and build a good relationship with the women. By taking this opportunity, neighbours report that an extra marital relation (10%), which mostly occurs to married women. The respondents report that the lack of access to public places for women (e.g., market and *bazar*) makes women dependent on men and vulnerable to manipulation and sexual exploitation (Fisher, 2010).

> Sometimes, neighbouring male members make comments and enter the house of the woman. In most cases, beautiful women face these types of problems more. (Male, 47, class XIV, In-depth individual interview)

Instances reported that, after cyclone *Aila*, men offered goods and services to vulnerable women such as a relief card for sex or engaging in relationships, which is reflected in the following study:

> Sometimes, neighbours, local political activists, UP chairman and UP members get engaged in illegal physical relationships with local women, particularly those who are more destitute. They lure these destitute women by offering extra facilities and benefits for relief and Kajer Binimoye Khaddo. Women become compelled to do as they are asked. Sometimes, women asked neighbours to buy something from bazar and comeback from the bazar with goods and take the opportunity. (Male, 42, class XVI, FGD with NGO activists)

If males do earn money, they send it to the family via the *bKash* system. This is the quickest and easiest approach to get money. However, if the remittance money does not come in time, the woman must manage the family by taking a loan from relatives and neighbours or buying goods from the nearby *bazar* on credit. Studies reported that, after leaving to earn money, certain migrated males did not send any money or sometimes the money was insufficient to meet the family's requirements.

Furthermore, financial hardship increases expectations from men for "dowry" type payments from wives' families to support the family in these new transitions. The increasing trend towards men abandoning their village families is causing considerable hardship for women and their natal families. Sometimes, males stop communication and remittances to the family. Women's vulnerability, marginalisation and insecurity is reflected in the following statement:

> My husband migrated to Gopalgonj to work and got married there. He then sent money but that is irregular and not enough. Often, I feel upset as while the other women have cooked meat in their house for their children, I cannot cook for my children. (Woman, 28, class IV, FGD with vulnerable women)

Another respondent reports on her vulnerability because of the migration of her husband as follows:

> My husband got married three times and it is been his luxury. Finally, after Aila, he married a woman who had certain money and land. My husband got married to woman to occupy her money and land. (Woman, 40, no schooling, FGD with vulnerable women)

4.4 Entitlements, Access to Resources and Women

When asked about resources/properties available to the family in the study area other than income earning and wage labour, respondents report that land (76%), cash money (14%), jewellery (24%), livestock and poultry (46%), shrimp *gher* (6%) and fishing equipment (22%) as properties/resources of the family. The properties and assets are categorised into productive and non-productive; the productive assets include livestock and poultry, shrimp *gher*, fishing equipment and land. The non-productive assets include jewellery and electronic goods. The most common property in the study area is land that includes cultivable agricultural land, homestead gardening/vegetable land, plot of house and gardening places around the residence. The majority of respondents (91%) report that they do not have entitlements and access to family properties and resources.

> I believe that it is about faith. A male does not trust a woman; therefore, women are lacking access to resources even women are lacking access to their own income. Trust is extremely much similar to religion; moreover, religion allows unequal rights for women to resources. (Woman, 45, class 12, in-depth individual interview)

Respondents report in the focus group that women feel inferior and alienated because they do not have rights/access to resources even to their jewellery. The status of women at every level of society is determined by the position of the husband or father. Although Islamic law gives a daughter the right to inherit 50% of a son's share of the patrimonial property, the right is extremely formal compared to real (Kabeer, 1990). Social norms are clearly violating these rights and denied women and girls land entitlement deeds, particularly in rural areas (Sayeed, 2010). Even when women do have property entitlements in the study area, it is extensively controlled by their husbands or male family members because women can neither cultivate their own land themselves nor can they easily negotiate directly with outsiders for labour and their needs. Although the situations have changed in recent years, particularly after *Aila*, as reported in this study.

A woman's dependency on men's income/wages led to a condition of vulnerability because it is not her own income and she has to seek permission if she wants to spend. Dependency on her husband's income and wages and permission to spend uphold lacking entitlements for women and create vulnerability for women. Sometimes, property ownership does not mean women are empowered if they do not have control on the return or income from the property. Therefore, property control and income from property and earning by women effectively empowers them most as respondents. Women who are from better off families will be empowered if they are earning themselves.

What are the constraints for ensuring property rights? Respondents report constraints that prevent women from entitlement to resources and these are all created by society that includes social norms (22%), patriarchal systems (37%), negligence to women (20%), women not aware of their rights (12%), no trust given to women (7%) and religious norms (11%). This study reports that when women lose assets such as jewellery, they have a weaker influence in the family because jewellery is the last resource women can sell for an extreme crisis. Accordingly, a woman respondent delineated her rights in the following manner:

> I had certain gold that I received from my parents' family; however, I do not know where my husband hides them. When I was required to occasionally wear them, I asked my husband and then he used to give them to me but sometimes he declined saying you do not need to wear that all the time. All of a sudden, after Aila, he informed me that the ornaments are missing. My husband said flood water washed them away because flood water entered the house and we were in the shelter centre for > 1 week. I became suspicious and have been

beaten several times by my husband while I asked him forcefully to give them back to me. (Woman, 42, class VI, FGD with vulnerable women)

More importantly, unequal gender relations in the study area affected income-sharing and resources at an extended family level. One woman describes the following:

All our land was under my father-in-law who transferred his land and properties to my husband before he passed away. Previously, my father-in-law held all properties in his name because of social and religious norms; moreover, males did not trust women. (Woman, 50, illiterate, FGD with vulnerable women)

Sen (1981b) reported that destitution in the Bengal famine was wide spread and massive because of the entitlement-related failure. Because of Sen's argument when the participants in this study were asked about women's rights to resources and the relation to vulnerability, 95% of respondents report entitlements/ownership and access to resources reducing vulnerability. The respondents confidently reported that in cases where women have access to resources, social practices and constraints are being relaxed or removed; moreover, the status of women will certainly be upheld and vulnerability will be reduced irrespective of whether a woman is in a wealthier or poorer family. In a wealthier family, if women have no rights and access to resources/properties, they will be vulnerable:

I had four cows and lots of poultry. I was used to earning around 1,000 taka on a weekly basis by selling milk and eggs. However, Aila washed away my four cows and all poultry as I did not have sufficient time to take care and take them on the beribadh. I managed and met many requirements of my family, and my husband even did not know that. I was solvent and self-reliant financially even after giving some of my income to my husband. I made many decisions by myself; however, after Aila, I was dependent for anything on my husband. (Woman, 47, can sign only, FGD with CPP members)

A woman thus reported the following:

I believe that if I were to have a bank balance of 50,000 taka, my husband will ask me for my opinion before making any decision. In rural areas, there is a proverb "bone without meat dogs even do not eat." If I have 5 bighas land, everybody will behave well with me and everybody will care for me. (Woman, 23, class XI, in-depth individual interview)

In terms of the association between vulnerability and entitlement/access to resources, one NGO activist during FGD opined as follows:

If you have no wealth and no rights to wealth, you will be much more vulnerable; if you have wealth and have entitlements to wealth, you will be less vulnerable. Accordingly, women are extremely vulnerable because they have no or limited rights to wealth and resources. Therefore, if we can ensure rights or access to resources of women, it will be better for families and for themselves. (Male, 48, class XIV, FGD with NGO activists)

Unequal gender relations, dependency of women and differentiation in coping mechanisms affected the capacity to use micro-credit loans that are considered in the woman's name from NGOs, *samity*, cooperatives and *mohajon.* The respondents report that when there are no fish in the river and no crops in the field, the people of Sutarkhali and Kali Nagor maintain the family by taking loans from NGOs and *samity* (23%) as well as from *mohajan* (24%). To survive during a crisis, women take loans in their name and then give the loan-related money to their husband. Sometimes, women even take a loan against their jewellery and women are liable to repay loan-related instalments but the credit of loans is then used by men. A woman respondent describes loan taking and control as follows:

> NGOs and other informal loan authorities grant loans in the name of women because women are there and they cannot go and flee away like men. Moreover, women have jewellery, poultry, and livestock to pay the loan-related instalments. However, men took control over the loan because they think that they can invest it in a better manner. (Woman, 37, class V, FGD with vulnerable women)

Women in the study area earn money by doing homestead gardening, which meets their household requirements and sometimes they sell in the local market. Women do forestation around their house by planting wood trees and fuel wood trees in coastal areas, which is not only for earning money but it also protects them from storms and strong winds during a disaster. The respondents argue that more homestead gardening and more forestation produces oxygen and reduces carbon emission.

Women view the requirement to work for income as a necessity to support their family; nevertheless, it is a necessity that is challenging customs that formerly constrained their lives (Alston, 2015). The respondents report that economic control is vested in the hands of men; furthermore, rural women seldom get support from their husbands. In a traditional patriarchal society such as Bangladesh, other studies report that freedom in cash expenditure opportunities for rural women remains quite limited (Afrin et al. 2008; Parveen, 2007; Sultana, 2010, 2011). Studies report that the division of labour, lack of rights to assets, and widespread and profound inequalities leave women more vulnerable and less able to cope with additional stress and deprivation brought about by climate change and during disasters (Reyes, 2002; Roy & Venema, 2002).

4.5 Entitlements, Involvement with Decision-Making Process and Women

Women's participation in decision-making is extremely limited (Bose et al., 2009; Hossain & Bose, 2004), although the participation of women at all levels of decision-making is required for their betterment and to achieve equality in the family and society (Hoque & Itohara, 2008). In Sutarkhali and Kali Nagor, decisions related to economic activities (money spent, any selling and buying of goods, involvement with earning activities), community activities (helping destitute people, voluntary road/dam construction work, helping neighbours on different occasions) and family ceremonies are generally considered by the heads of households who are predominantly (95%) men.

Respondents report that a small number of women have opportunities to independently make decisions; because of the lack of decision-making opportunities, certain women cannot get involved with income-generating activities. The incidents of mutual decisions were limited. Normally, after making a decision, the head of the household discusses it with other family members but the respondents report that it does not matter whether the women members agree or disagree. The decision taken by the head of the household takes is final. Women have to take permissions from their husband or from male family members prior to taking decisions, which is identified as a problem by respondents. A woman respondent portrayed the role of women in decision-making as follows:

> My husband makes all decisions in the family; he does not ask me before making any decision and does not even discuss with others. Sometimes, I can speak out after listening to the decision, but my husband makes the final decision. (Woman, 35, class X, FGD with vulnerable women)

Similarly, other woman shared her experience and reported the following:

> Whatever I want to do, I will first ask my husband. If he agrees and permits me to do those particular activities, only then can I do them. Without taking prior permission from my husband, I do not do any activities; it creates problems and chaos in the family. (Woman, 32, class IX, FGD with vulnerable women)

Survey data reveal that when making decisions for any purpose in the family, women in most cases cannot play an equal role in the decision-making process only because they are women. Respondents report that the role of women in the decision-making process in the family ranges from 1 to 100% in the study area. Most decisions are taken by women in relation to home gardening, cattle, goat and poultry rearing and post-harvesting activities such as seed production and storage. Among 263 respondents of the study, 34% (89) of respondents play a role in decision-making for 40%–59% cases in the family and 22% (58) of the respondents play a role in decision-making for 20%–39% cases in the family compared to 9.9% (26) respondents playing a role in decision-making in the family for 80%–100% of cases. Interestingly, 24% (62) of the respondents report that they have no access to the decision-making process (Fig. 4.2) and most of these women are from better off families who do not require working outside for income and are not allowed to either as per the respondents.

The majority of respondents both men and women agree that lack of access/entitlements to resources and properties, and lack of control over income and loans affect women's self-esteem. A woman respondent reports the following:

> I know how the household is managed and my experiences help develop coping mechanisms during a disastrous but I cannot independently take the tasks. I feel it is too complex and it is too difficult to make any decision without the opinion of male members. Moreover, I do that because I am born and brought up with this sort of belief. (Woman, 45, class V, FGD with PIC)

However, in times of change, this does not work extremely well. The respondents argue that during a disaster and normal times, women can manage the worse situation

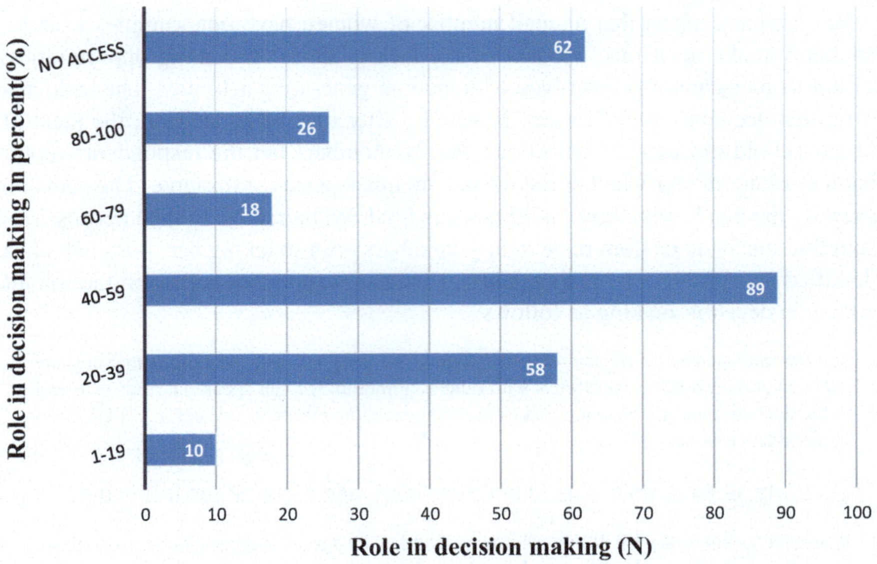

Fig. 4.2 Distribution of respondents' role in decision-making (in percent)

of the family better if they have access to resources, income of the family and micro-credit loans, as well as if they could have involvement with the decision-making process. In answer to this question, do you think access to resources impacted the decision-making capacity of women and how the respondents argue that the family would be better off if women had access to decision-making process. Access and entitlement provides freedom to women, which makes them confident to make decisions on their own and are more capable to manage households and related activities. This helps them share with their husband whatever ideas they have in mind. The entitlements and decision-making capacity and role of women in coastal areas are reflected by a man and woman as follows:

> Definitely access to resources has an impact on decision-making capacity and level of vulnerability of the women. If we suppose women own their income and if they have access to their husband's income, they can make a decision for the betterment of the family. In particular, in a disastrous time, for the lack of money, women cannot hire a van to go into the shelter centre. Sharing between man and woman leads to good results. (Male, 47, class XIV, In-depth individual interview)

> We talk with each other before making any decisions. I used to request my husband for buying clothes for my kids, but I do not need to tell my husband. Now, attitudes have changed; moreover, I got assistance and support from my father-in- law. (Women, 39, class IX, FGD with vulnerable women)

The level of education and jobs play an important role in empowering women; more education helps women to have access and rights to resources and more participation and to some extent control over decision-making. In the study area, there were certain exceptions in terms of access to resources, rights of properties and control over income and loans, although it does not reflect the general scenario of the coastal community. A woman reports the following:

> I have equal rights and access to resources as my husband. However, it is not the reflection of the whole society and it is not common practice. It depends on education, nature of jobs, and trust and mutual understanding of each other. I have freedom and my husband has given me independence to spend earned money when I want. (Woman, 30, class IX, FGD with vulnerable women)

One knowledgeable young woman who is a primary school teacher explained her status as follows:

> I am a teacher in a local primary school and my husband migrated to Saudi Arabia. All decisions in my family are made by me. I do not wait for my husband's remittance, although sometimes I face difficulties to maintain my family. I think I can solve my problems by myself. (Woman, 35, class XIV, FGD with UDMC)

Husbands take care of these women who have scope to get money and properties from their parents, particularly if the wife's father is rich and does not have a son or male to inherit.

> My father-in-law and husband did not take even a piece of cotton from my father; however, there are some people who expect benefits. Taking dowry, financial support, and other benefits from wife's parents is a psychological matter of the head of the family. To obtain dowry, certain families arrange early marriage for the son. Some husbands force their wives to get money from her parent's house for business. A wife gets good behaviour from her in laws if she can get money and she plays a role in decision-making. These issues happen only if there is her own money and own property. (Woman, 23, class XI, in-depth individual interview)

Respect, inclusion and better treatment by their husbands in decision-making because of their access/entitlement to properties is a process of women empowerment. Access/entitlement to resources places women in a stronger position and makes women capable of reducing their dependency and powerlessness condition on their husband and they gain control of their lives.

In a FGD with vulnerable women, discussants report that, if their decisions are ignored in the family, they are ignored in the *samity* meeting. The respondents report that when women are sometimes called to attend the meeting of the *samity*, husbands prevent women from attending the meeting, and sometimes husbands attend the meeting instead. In terms of valuing women's opinions in the meeting, women respondents report that only if the women's opinion is similar to that of the men will it be accepted; otherwise, the opinion of women is ignored. From the FGDs with coastal people, it is basically found that the women who are senior in the family such as mothers, women with a job, educated, heads of households and women whose husbands have migrated are able to play a role in the decision-making process in the family. One aged woman discussant reports the following:

> My son earns and does carry the family expenses. He does all activities, but he asks for my opinion and shares with me as mother and senior member of the family before doing anything for the family. (Woman, 62, no schooling, FGD with female-headed households)

Respondents in the study area believe that decision-making and bargaining power of women will increase and ensure in the family and in society only if entitlements/access to resources for women are ensured and increased. Agarwal, as cited in Enarson (2000), reports resources and decision-making power, which leaves men in a stronger "fall-back position". Previously, studies (Alston, 2015; Haider & Akhter, 1999; Miller, 2001; Saha, 2015) describe that economic matters such as land acquisition, selling, buying, crop production, marketing, sale or pawn of valuable goods and animals, involvement with income-generating activities both within households or outside and initiatives/investment for business such as livestock and poultry were decided by men. However, the situation and dynamics of decision-making power of women has changed/improved in the last few years, particularly after *Aila*, although the experiences and participation of women in the decision-making process are mixed and contradictory because development/changes in the coastal area are still incomplete and ongoing. This study shows that certain women with entitlements to resources in Sutarkhali and Kali Nagor were successful and make their decisions in the family and they play a role in decision-making at a community level. For compelling society to hear the voices of women in the Bangladeshi coastal area, ensuring access, right and control to resources/properties is a fundamental/pivotal factor.

4.6 Skill and Knowledge-Based Training and Role in Decision-Making

While GOs are engaged in disaster preparation and early warning strategies, NGOs possibly participate in the recovery phase of a disaster and NGOs have had an important role in reducing the impacts of hazards on villages (Alston, 2015). The degree of vulnerability and risks caused in the study area because of *Aila* have been reduced by NGOs after considering different measures, and most measures were local skill and knowledge-based. This study shows that the degree of vulnerability and risks could be managed better by involving women at a large scale in the coastal area. The respondents report that social programmes such as training improved women's capacity to face disasters. This study examines a positive association between various training programmes for women undertaken by NGOs and reduction of vulnerability. An NGO activist describes the following:

> As per the policy, NGOs target 70%-80% of women to be included as participants in the training and other programs in the coastal area. Hindu–Muslim is not an issue to be concerned about; women rather are counted as the beneficiary and NGOs focus on women, thus increasing their capability using their local skills and knowledge. For example, a woman with a disabled husband is our target beneficiary for mati kata, a poultry and homestead gardening project with an aim to make women economically self-reliant. (Male, 25, class XIV, In-depth individual interview)

People in coastal areas believe vulnerability during a disaster would have been managed better if they still had training by NGOs such as those they had after *Aila*.

Alston (2015) reported that NGOs are working across Bangladesh and providing ongoing assistance to affected communities; moreover, their contribution is recognised by people in the community. Many are aware that NGOs have saved their lives. After *Aila*, 68% of survey respondents participated in training provided by NGOs.

The respondents report that women require to be financially solvent and self-reliant and that will happen only when women earn money. The respondents argue that if women can earn money and if women get involved with income-generating activities, coastal women would be empowered and would be capable of reducing vulnerability by themselves. Vulnerable women agree that training provided by NGOs on different issues is making women more confident. Training provided by NGOs helps women in the study area in empowerment more and better than getting food and relief because it gives them the opportunity to earn money, and using these experiences of training women are becoming more self-dependent.

A focus group respondent and NGO activist initially reports that the participation rate of women in training programmes was low but on a daily basis, the participation is increasing because of positive changes in practices in the community. Because of training, the abandoned land surrounding residences is now under vegetable gardening, information dissemination is more effective, people have become self-sufficient and poverty is being managed better. NGOs in the study area primarily focus on income-generating activities such as training on poultry and fisheries (41%), sewing (16%), vegetable gardening and forestation (24%), livestock rearing (19%) and disaster preparedness (48%). Respondents report that in their locality, *samity* was formed by an NGO comprising 35 women members who cultivate sunflowers with the NGO's assistance. A *samity* formed by women is not always welcomed by the community and, to some extent, family members are not welcome either. However, in terms of the obstructions to forming a *samity*, a woman respondent reports the following:

> I was initially obstructed by my husband and later on by other people. My husband always discouraged me from doing samity. Sometimes, involvement with samity leads to chaos and unhappiness in the family. I do samity by concealing it from my husband and disclose it only when I make a profit. (Woman, 25, class XIV, FGD with vulnerable women)

Money will lead to the emancipation of women and will help their role in the decision-making process. Accordingly, one woman respondent described the following:

> If women family members were not getting involved with earning activities, the family might have to starve. However, women are now earning and thus family members do not have to starve. Moreover, there was previously no scope for empowering women. In our patriarchal society, we [women] had to do whatever men wanted. However, since then [after Aila], attending different training provided by NGOs now I can make an income. I am asked to provide my opinion before taking any family-related decision. (Woman, 25, class XIV, FGD with vulnerable women)

Alston (2015) reported that the empowerment of women such as taking a role in decision-making is happening in the Bangladeshi coastal area particularly after *Aila* because women have the opportunity to get involved with earning activities that have been largely promoted by NGOs. Respondents argue that husbands cannot make decisions alone. They must focus on their wives too. If solvency comes by women's effort, the husband will obviously give them priority because both are contributing to the family's well-being. One male respondent adds the following:

> Women are empowered because of GO and NGO aid, and programmes and training are completed in the name of women. As the right of women is a priority of GO and NGOs, husbands in the family realise there should be a common understanding for the betterment of the family. (Male, 60, class X, FGD with vulnerable community people)

In this study, 97% of survey respondents add that different training programmes are reducing vulnerability and improving the role and responsibilities of women in the family and at the community level. Different training programmes provided by NGOs created an opportunity for women in the study area to take multiple training options, including disaster preparedness, poultry and fisheries, homestead gardening, livestock rearing, sewing and small business. This training creates opportunities for income for women in coastal areas and training has empowered women by increasing income (66%), reducing poverty (52%), increasing food security (44%), increasing their role in decision-making (32%), ensuring hygienic sanitation (25.5%) and ensuring schooling and education for children (14%).

The participation of women in different training created income opportunities for women in the study area, which boosted the capability of women such as expenditure in the family. This indicates increase in income is a tool/component for women to be equipped to encounter poverty or for reducing the poverty level because it has improved the material quality of life such as food habits, number of food intakes in a day and education of children.

The study reports that after attending training programmes, women are getting involved with earning activities and they are working outside with males; this has happened because of attitudinal changes in the coastal area in other ways because of NGOs' programmes. Traditionally, in Bangladesh, a good woman is not expected to "roam the streets" but to stay at home with her husband, children and female family members (Panday, 2016). However, the findings of the study show that training provided by NGOs at a rural level resulted in women earning money and being self-dependent to go out, which ultimately remarkably contributed to the family and community's well-being. In other words, women achieved more respect from family members, more freedom to go out of the house or villages, and increased effectiveness in decision-making (Panday, 2016).

4.7 Summary of the Chapter and CAM Model

The above discussions in this chapter conclude that socio-economic and cultural challenges influence the disproportionate vulnerability of women, changes in degrees of vulnerability and coping and adaptation mechanisms of women in the coastal area in Bangladesh. These socio-economic and cultural challenges/pressures include limited access to income-generating activities, lack of entitlements/access and control of resources, and less participation in decision-making process and migration of males.

For example, because of the position of the father and husband in Muslim religious law (which dominates GO law about the inheritance of properties), the inheritance of the property by daughters is different for sons. Whatever the amount of property that daughters inherit, husbands or other male members of the family take over the control of the property and women have limited opportunity to negotiate with outsiders. While access to and control of resources, involvement of women with diverse income-generating activities, and access to family income creates an opportunity to improve the socio-economic conditions for women. The control of wealth promotes participation in the decision-making process and ensuring income. The limited access of women to income-generating activities, lack of control of resources and migration of male family members are pressures that increase vulnerability during a disaster for women and for their households. As a theory, PAR directs attention to pressures that come from the existing social structure with its considerable challenges to women for improving their coping mechanisms.

Migration (temporary and permanent) of male family members left women alone to manage dependents in the family; this has an association with social and sexual insecurity, in addition to extra responsibility. In the absence of male members and lack of access to public places such as local market (*bazar*), women become reliant on male neighbours. However, this situation is worse for female-headed households during a disaster. For instance, women are discriminated against and get lower wages, less aid and relief. They are being sexually harassed, discriminated against by attitudes of society, and suffer from a lack of ties with relatives in the absence of a male protector and guardian in the family.

Compared to women with better economic conditions, women with poor economic backgrounds and female-headed households in the society are more prone than others to damage, loss and suffering. The policies of GO to prioritise women particularly gave a priority to under-privileged and women from poor socio-economic backgrounds. Policies that address social safety, and training and awareness programmes have brought women out of the home, which ultimately reduce their vulnerability. Changes in societal, patriarchal, religious attitudes, economic hardship, limited employment and earning opportunities immediately after *Aila* forced women to remain involved with natural resource management, outdoor work, homestead gardening, poultry and livestock, thus improving access to resources for women as reported by the access theory and entitlement approach. The empirical evidence shows that the increased access to the available resources in the coastal area led to a positive outcome.

The process of changes in the study area (as a consequence of appropriate, science-based and convenient training relevant to local practices) and involvement of women and female-headed households with different awareness programmes assist women to stop using their less effective methods of self-protection, coping and adaptation during a disaster. To some extent, policy-makers, development planners, state authority and international agencies influenced by PAR require to reconsider the less obvious vulnerable, such as women from wealthier families who suffer, particularly psychologically, during a financial crisis. They lack certain coping skills of poor women and their well-known identity as wealthier in society creates coping problems.

This study argues that despite unsystematic and inadequate formal adaptive measures to encounter challenges during a disaster, women in the coastal area take various measures for diversified challenges, including employment opportunities, entitlements to resources and decisions. The measures taken by women require effort. Consequently, women have developed their skills, knowledge and capacity and have started earning money, thus involving themselves with productive works both in household and outside. This leads to social change and social transformation in the coastal area (Fig. 4.3).

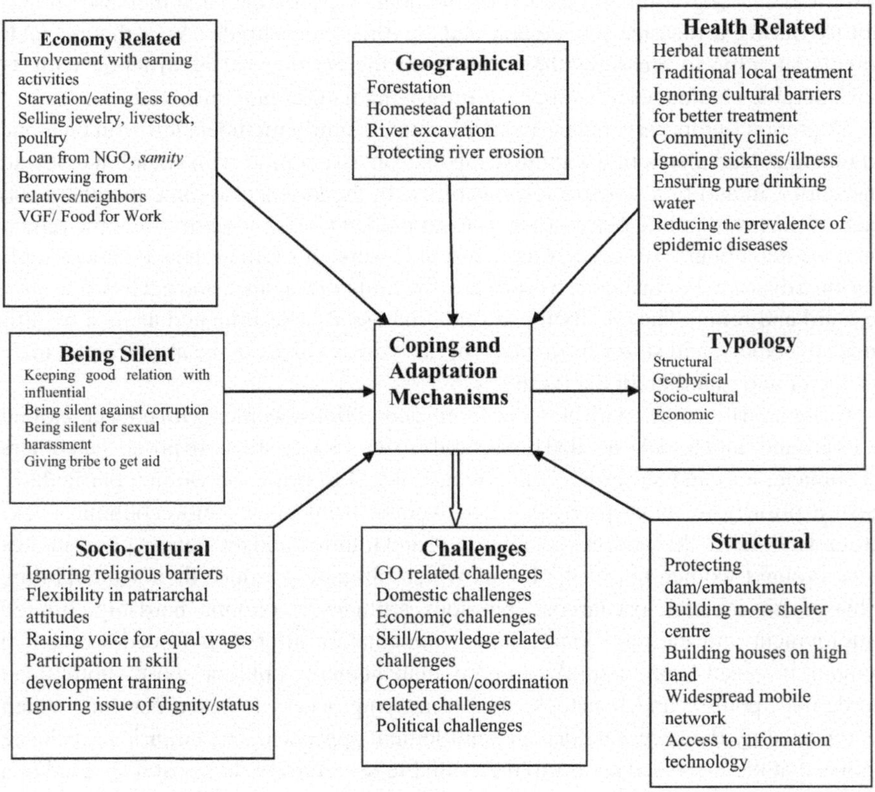

Fig. 4.3 Coping and adaptation model (CAM)

Chapter 5
Women's Empowerment, Role of GO and NGO in Disaster Management

5.1 Introduction

Societal development is not just an economic change; rather, it is good governance and ensuring participation of all groups for the benefits of socio-cultural change. GO and NGOs as actors contribute to empowerment and developing the agency of women in coastal areas. Socio-cultural development has promoted the freedom of women to independently act and be involved with socio-economic and political opportunities (empowerment) and capacity to control their lives and capacity to raise voice for their rights (agency) in coastal areas. Empowerment and agency of women in coastal Bangladesh are reflected in their access to public places such as local *bazar,* representation in different committee meetings and raising issues, participation in the election process and involvement with GO and NGOs for the effective implementation of projects. This reduces their dependency, increases their awareness, develops their bargaining capacity and improves their leadership for management at both domestic and community level. They allow women to talk about their social, political and economic rights and push them out in public platforms. However, GO is unsuccessful to certain extent in ensuring their safety programmes to empower women because of the lack of good governance, bias and the complex authoritarian structure of administration and lack of access to the individual at a grassroots level.

NGOs have picked up on this point and ensured different programmes for women such that they can control their lives and they can have capacity to make decisions. NGOs created opportunity for participating in local elections and for involvement with project implementation, which has resulted in changes in social structure and social value including at both family and community level. The social and cultural grounding in which women's empowerment takes place is initiated and nurtured in a better manner by NGOs prioritising and mainstreaming marginalised people at a grassroots level. Examples are creating micro-financing opportunities for women, capacity-building programmes for women, prioritising the interests of the individual and increasing the interests of the underprivileged. NGOs deal with stakeholders at a grassroots level and consequently they reach and understand individuals to know their

M. M. Khan, *Disaster and Gender in Coastal Bangladesh*,
https://doi.org/10.1007/978-981-19-3284-7_5

desires despite challenges from societal and infrastructural facilities and unfriendly collaboration of GO. Women empowerment as a catalyst ensures sustainable development because it brings structural changes, e.g., it challenges patriarchal authority and existing social fabric such as attitudes about women's mobility outside the home. This chapter addresses the indicators of empowerment and agency of women, role of GO and NGOs in empowering women, why NGOs are more successful in empowering women and remaining challenges in empowering women in the coastal area.

5.2 Indicators of Women Empowerment

The concept of empowerment has different interpretation in meaning and understanding in different socio-cultural and political contexts. In this study, the empowerment of women is used to indicate the involvement and participation of women with the opportunities and platforms available in the community to work for the betterment of the community and participation and involvement of women with awareness and capacity-building programmes (Kabeer, 2000, 1994; Mahmud et al., 2014; Oxfam, 1995; Panday, 2016). In the Bangladeshi coastal areas, both GO and NGOs are committed to empowering women by ensuring and fostering the participation of women in politics along with socio-cultural, economic, psycho-social and awareness programmes. Consequently, changes started to happen in socio-cultural, economic and political spheres to ensure the psycho-social well-being of women (Jain, 2017; Mahmud et al., 2014, 2017; Panday, 2016). These changes played a role as a catalyst for women to gain access to public platforms of social power where decisions are made for society.

Social power indicates the power that places an individual or an agent in a position within a social relationship to perform their will despite challenges where individual or agents cannot be independently considered; rather, they are interdependent because everyone has power and ability to carry out their will (Fiske & Berdahl, 2007). In this study, social power indicates the power that creates the ability of women to achieve goals, has control over resources and control over what people want, although there are challenges to achieving these aims. Moreover, this social power of women is manifested in the coastal society because of a multitude of social process and multi-social agents (Fiske & Berdahl, 2007). In this study, the social power of women in nature is flexible. This flexibility of social power is accredited/acknowledged by institutions. The intervention of different agents such as GO and NGO in the lives of women includes economic and political participation, socio-cultural flexibility and changes in psycho-social status.

In the coastal area, access of women to social power increases the ability of women to independently work and enhances their ability to influence any decisions taken by committees. For example, the economic participation of women in the coastal area includes women in food for work, cash for work and poultry and livestock-rearing programmes. Women participate in other available income-generating activities that ensure their access to and control over income and family resources. Moreover, they have access to local markets and businesses. Moreover, economic participation of women in the study area is now aware of politics. Consequently, women's representations in committees, local GOs and other decision-making platforms are evident, which makes their voice heard by the community. "Getting voice heard" of women by the community beyond their families helps women gain access to public areas where decisions are made. The access and greater access of women to those platforms enhance the capacity of the individual to independently make choices, e.g., children's education.

Furthermore, women start to identify the causes of their powerlessness and disproportionate experience of vulnerability, which has helped women raise their voices for their rights. Many studies reported that identifying the constraints that lead to their powerlessness encouraged women to take control of their lives, which in turn started to lead to changes in the conditions of their lives (Mahmud, 2010; Panday, 2016). The realisation of constraints leading to powerlessness is the first step to being empowered for women. This study addresses raising self-awareness, identifying the challenges and capacity building for actions to do independently as empowerment. For example, the challenges identified by women when they are kept away from involvement in GO projects. Moreover, they know their experiences are ignored during the implementation of projects. Because of awareness of the male family members and community, the religious practices and male attitudes towards women have become more flexible. As a reflection of women empowerment, this study shows that women make decisions on the education of their children, become involved with extra familial groups and programmes and build social networks, which creates opportunities for them to get access to public and social spaces. Participation and awareness built their capacity and developed agency about legal rights and legal support by GO-NGO against violence. Awareness contributed to the women's self-esteem, psycho-social well-being and collective awareness of the community to allow women to be part of development and community mobilisation for women's empowerment. Therefore, women empowerment in the coastal area broadly addresses multi-dimensional aspects, including socio-cultural, economic, psycho-social, political and familial (Mahmud et al., 2017; Panday, 2017), which have specific aspects of contribution for women's empowerment. (This has been summarised in Table 5.1).

5.3 Representation of Women in Local Institutions and Committees

Public platforms indicate different committees such as Union Disaster Management Committee (UDMC), Project Implementation Committee (PIC), Cyclone Preparedness Programme (CPP), *Bazar* Committee, and School Managing Committee (SMC) of the study area where different decisions are made in the community's interests. For example, UDMC is permanently operating in the study area and has more involvement with the people of the community than other committees in the community such as PIC. UDMC ensures the representatives of the people of the community are on the committee, thus establishing networking with GO, tapping financial and technical support from GO and other organisations, mobilising the community at large, facilitating social, economic and physical rehabilitation of community, coordinating with GO and aid agencies to receive assistance in rehabilitation. They share damage information with GO and other agencies, thus ensuring integration of risk reduction measures at the reconstruction and rehabilitation phase, and developing strategies for future improvements to cope with disaster risks. Because of activities throughout the year, UDMC can influence the lives of the people of the community. There is an annual budget of two *lakh* taka, which is not a big amount. UDMC is the only committee that gets this budget. They make decisions about how to manage resources and aid during and after a disaster.

Only PIC looks after the implementation of any community projects, including post-disaster and regular projects at normal times, the *bazar* committee deals with any problems related to local *bazar* and *bazar* management, including toll, tax and setting up new shops. The SMC makes decisions about overall management, including setting fees. The researcher had an expectation that all respondents would know about different committees in their locality; however, the study results show that 77% of survey respondents said "yes" they know about these different committees.

For example, the involvement of women with the different disaster management committees has been ensured by the Disaster Management Act 2012 and GoB, (2010a) with the aim of creating space for women's participation and giving them a voice in the decision-making processes for an effective solution and addressing the challenges that affect their lives (Panday, 2016). Despite the instruction for women to be involved, field data show women are not always included and cannot play their role as per acts argued by respondents. Both formal and informal committees are formed by elected local representatives such as UP chairman, UP members and local political leaders with the assistance and coordination of the *upazila* chairman and UNO.

In response to the question that whether women have any problem working with different committees in the community, it is argued by respondents that women on committees face limitations. This has been reflected by a female member of PIC committee who describes her experience in terms other participation in PIC meetings: *I am a member of PIC committee but I hardly attend any meetings; instead, my husband attends for me* (Woman, 45, class V, FGD with PIC members).

Why a husband attends the PIC meeting rather than his wife, a male member of the PIC committee explains (and all other members agree) the following:

> Women cannot provide sufficient time to monitor different activities in the locality but a male, and her husband provides time to the committee and a male member can attend to committee business at any time and any place. Sometimes, we sit down for a meeting in the evening and in the night at the community club [a social club of the village where villagers get together during their leisure time, village meetings/arbitration takes place there] she can not attend because she has to maintain her household, including livestock, poultry and caring for other members and has to arrange dinner for family members. Sometimes, we have to go for meetings with GO officials in Sadar, which is 20km away and it takes 1.5–2h and we are late getting back to the village. To summarize, there are certain rural issues and norms

> for women that are discouraging women from going out any time. If they leave the house, who will take care of their households' activities? Rather than a woman member's opinion in the meeting, we take the opinion of her husband and the husband discusses this with his wife. (Male, 38, class XII, FGD with PIC members)

However, the respondents of the study area believe that times have changed and the role and position of women, including women from better off families have improved in multiple aspects of women's lives, both personal and public; this is noted in other studies (Kabeer, 2011). In this study, being involved with public affairs includes the involvement of women with different committees, freedom to go out, women's leadership in disaster management, roles in decision-making and direct involvement with politics. In terms of changing roles, involvement in public community affairs, and the responsibilities of women in the locality for different activities and programmes, the respondents described the cyclone *Aila* as a milestone for the coastal community people. This is because women now have a sense of self-worth and social identity and their ability to question and capacity to exercise and take control over their own lives (Kabeer, 2011). This study shows that most importantly, before *Aila* women from the local community did not know about different committees and their involvement with committees; however, women are now well aware of this. In terms of getting involved with committees, a female member describes involvement with CPP as follows:

> Before Aila, CPP was not in our community and was formed after Aila. CPP disseminates early warning before any disaster in the local area through a hand mike [a loud speaker run by battery and convenient to carry]. We [CPP members] help people take their poultry, cattle and necessary goods to a safe shelter. (Woman, 36, class IX, FGD with CPP members)

Of the 263 respondents, 68% of respondents knew there are women members in different committees; however, a number of respondents (25%) reported that they "don't know" whether there is involvement of women members. These respondents (particularly women from better off or mostly conservative families) are more limited to working in their households and absolutely dependent on male members of their households. When asked about the participation rate of women in the meetings of the committee, the respondents report that the number of participants and active involvement of women has rapidly increased compared to the past. Women attend meetings and women sometimes ask the head of committees about the next meeting time and agenda of the meeting by themselves. Because of the participation of women in rural areas of Bangladesh in the decision-making processes of UPs, they were enabled to raise their voice more effectively (Panday, 2016). The information of respondents is positively reflected by the respondents of the survey where 60% of respondents said that "yes" women members participate in meetings and different programmes organised by committees, where as 61% of respondents say that women do not face any obstruction in attending meetings. However, 40% of respondents report that the degree of attendance of women in the meetings is affected by different social discriminatory norms such as social norms, religious norms/*purdha* systems, patriarchal system, transport and communication, and political interference. Identifying

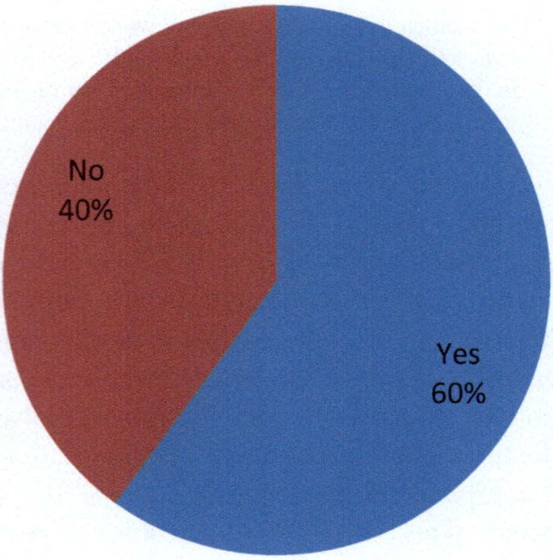

Fig. 5.1 Distribution of opinion on whether women participate in different committee meetings

the challenges by respondents (40%) reports their power and ability to identify challenges over conflict and confrontation with the existing social system and powerful people, which relates to their ability to argue for women empowerment (Panday, 2016) (Fig. 5.1).

> Ensuring the participation of women at different meetings of different committees is definitely good. Women gather new knowledge about problems that they did not know about by attending the meeting, e.g., the importance of education for children and rights of the mother and caring tips for pregnant women. Nowadays, I do not hear of any problems and obstruction for women attending meetings. Situation has now improved; I am a member of the school managing committee, and my husband inspires me to attend the school meeting. (Woman, 45, class XII, in-depth individual interview)

In response to being asked about changes and improvements at a community level, another woman respondent reports the following:

> All programmes and projects in our locality are primarily for raising awareness. If someone repeatedly attends different programmes and knows about responsibilities and rights of women, there will be definitely a change in their thinking. The involvement of women with different committees and training has helped to remove superstition from society. (Woman, 23, class XI, in-depth individual interview)

To examine the association between having more involvement of women with various committees and whether women empowerment and the effectiveness of management was occurring, this study reported that involvement is effective because women have easy access to other women to increase their awareness (30%). Moreover, they have easy access inside households and to women (46%), and it is easy to ensure more social security for women (10%), easy to enhance women's capacity (9%), women have better local knowledge (7%), and women are good household managers (5%). These factors make it easier for women and make them confident to be involved with different committees to efficiently manage their lives and to negotiate better terms in their relationship with others. This increases their ability to participate on equal terms with men in reshaping society to better accord (Kabeer, 2011). Field data report that 87% respondents said "yes" (Fig. 5.2), and the involvement of women with different committee reduces their powerlessness and vulnerability and speeds up effective and efficient social management. For example, when asked how more efficient disaster management is possible by involving more women, male respondents report that during a disastrous there are certain situations where only women can get access to deal with problems. A male during a FGD with CPP members describes the following:

> Imagine the situation of the shelter centre during a disaster. In the shelter centre, if women members are deployed, they can they manage better because women have easy access to women and women can solve their problems best. (Male, 25, class VIII, FGD with CPP)

Another male respondent adds that certainly:

> There are certain problems that cannot be solved by men. Men can move easily but do not have access everywhere, e.g., if there is a pregnant woman only a woman can help. A man can not go there, not even the husband during birth, because of our socio-cultural practices. (Male, 34, class XII, FGD with PIC members)

> It is positive to involve more women in the disaster management process. However, it is not necessarily right that if the number of women members increases in the disaster management committee, it will be beneficial. What we require to ensure is the additional involvement of women, e.g., village-para-based committee and programmes can be formed for women only rather than being based in union headquarters. (Male, 38, class XII, FGD with UDMC)

Despite this optimism, studies show not only are social norms still significantly obstructing women from attending meetings in the study area but also political interference sometimes makes women refrain from attending meetings. Political corruption and politicisation of everything, including committees by political leaders, are resulting in underperformance and inefficiency of more powerful UDMC committees (TIB, 2015; Wickberg, 2012). The politicisation problem is described by a woman as follows:

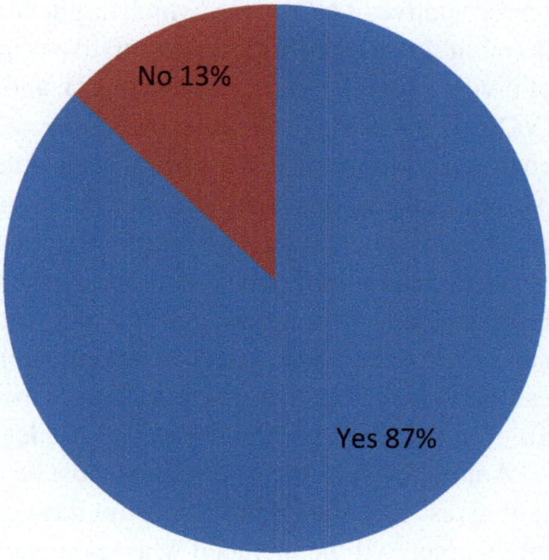

Fig. 5.2 Distribution of opinion regarding the involvement of women with different committees and reducing vulnerability and powerlessness

> I want to attend meetings but the problem is my political ideology is different from the majority of the members of the committee, including the president of the committee. Whatever I say is ignored just because of political ideology; therefore, I do not feel like attending meetings anymore and they do not want me to attend. Other committee members who are like minded to the president (with same political ideology) think that I attend meetings to create problems. (Woman, 33, class X, FGD with vulnerable women)

During discussion with female-headed households, vulnerable women described their experiences and said that after *Aila*, as part of the GO support programme, certain houses were constructed in their locality and this was monitored by PIC members. They report that the structure of houses was faulty and materials used were of extremely low quality; a female member of PIC committee raised this but was ignored. She replied in an interview as follows:

> We could not say anything; if we wanted to say anything we were told that we are getting things free of cost. So, we do not have the right to say anything and we accept low quality and offer a bribe sometimes. (Woman, 45, class V, FGD with PIC members)

People in coastal areas believe women can play an important role and women are seen to be proactive in emergencies. The proactive nature of women is described by certain discussants of CPP and their role in disastrous times is to make flags fly in public places and on the banks of the river to warn of disasters. When team leaders get news from the radio, TV or other sources, they inform the members. The members of CPP disseminate the information among people in the community. CPP leaders have the contact number of the weather forecast office and can get early information regarding cyclones and other hazards from there. Women as team members go door to door and force women to go in the shelter centre because they have easy access to other women as argued by respondents and reported in the study of Oxfam (2011).

However, the UDMC, according to SOD (2010) is headed by the chairman of the union *parishad* and must comprise 36 total members plus three members that the UP chairman can co-opt. Members in the UDMC will be included in the committee from all sectors of society, including UP members (12), teacher (1), GO officials (7), vulnerable women (2), representative of CPP (1), representative of Red Crescent (1), representative of NGOs (3) (which might be an NGO that is local, national or international in scope category), representative of peasant and fisherman (2), representative of civil society (3), religious leader (2), and representative of Village Defence Party (VDP) (1). The roles and responsibilities of UDMC include both risk reduction and emergency response at the warning period during the disaster and in the post-disaster period. The responsibility for risk reduction maintains local people informed and capable of taking practical measures to reduce risk at household and community level by arranging training and workshops, building capacity at individual and institutional level, and ensuring drinking water and life-saving medicines. Nevertheless, emergency response addresses coordinating relief activities, helping people to transfer essential resources to a safe place, and ensuring basic treatment for injured people. To perform all these effectively, GO has legislated for women's direct involvement.

Among UDMC members, there must be at least five women, including two representatives from the community who have experienced disasters, and more community women will be included who know the community and can talk for community women (civil society), in addition to three elected UP women members. The female members of the UDMC attend meetings and workshops on topics such as community vulnerability and risk assessment process and risks and risk reduction action plans. Female members of the TDMC (Thana Disaster Management Committee, one step higher committee from UDMC) and UDMC ensure participation from different vulnerable groups from the grassroots level and facilitate work of NGOs and assist the union chairman to mobilise resources for implementing the risk reduction action plan.

This study reveals from the documents provided by the union *parishad* office that Sutarkhali Union Disaster Management Committee formed in 2014–2015 has only 36 members, and the three extra members recommended under the Act are not being co-opted by the chairman. There are only five women on the committee, which is 14% of the total members; three of them are included by default as elected union *Parishad* women members. Only one is included on the committee as a representative of vulnerable women from the community; she is nominated by the chairman. A result of this study is that a maximum of eight UDMC meetings were held in 2014 and only three meetings in 2011. The lowest number of members present at a meeting in the last four years was ten (three women) in 2014, and the highest presence was 34 (10 women present) in 2011. The average number of members between year 2011 and 2014 in a meeting was 13 men and 3 women. Generally, three female members were present at all meetings with the exception of a meeting in 2011. Interestingly, the study revealed that the higher attendance in 2011 was attributed to a high volume of aid, relief and development projects in the coastal area after the *Sidr* (2007) and *Aila* (2009) periods, as argued by respondents. Subsequently, the presence and activism of the members of UDMC committee declined because there have been less relief, aid and projects. Certain projects that take a long time merge into projects for the next disaster.

The UDMC committee in Kamarkhola union has been formed by 39 members; nine of them were female members, which is 23% of the total number of committee. By default, three of them are included in the committee as elected members of the union, and another six female members represent vulnerable and disabled women, NGO and CPP members. The number of female members in Kamarkhola UDMC is higher compared to Sutarkhali where it is 13.9%. However, the summarised information and statistics about UDMC provided by the Kamarkhola union office revealed the poor representation of women in the committee meetings compared to male members. Documents provided by the Union office demonstrated that in 2011 there was only one UDMC meeting, and five UDMC meetings were consecutively conducted from 2013 and 2014. Throughout the years (from 2011 to 2014), the highest number of members present in a meeting was 36 (9 were women), the lowest was only four (no woman was present) and the average presence of members was 21. The average presence for men was 16, whereas that for women was five (5).

Respondents in the focus group with vulnerable women, female-headed households and community people stress that the number and representation of female members in UDMC is poor. Respondents in the focus group report that if women as members are welcomed more and allowed to freely work; if the numbers of women in the committee are increased, vulnerability during a disaster of the community will be reduced. Regarding the under-representation of women, Alston (2015) argued that because there are extremely few women in decision-making relating to climate change, there is a significant lack of understanding relating to the gendered nuances of climate impact, which can lead to an unintended reinforcement of gender inequalities by practices such as inequitable access to resources, aid, reconstruction work and GO practices. In terms of under-representation and inequitable access to resources, a woman respondent described the following:

> Though the government allocates construction materials, cash money, and 20 kg of rice in a month for disaster victims; as part of the emergency relief programme I received less aid and resources just because I am a widow and live alone. (Woman, 49, class VI, FGD with female-headed household)

The problems and obstructions to the inclusion of women in committees and programmes are fabricated or exaggerated by the male members of the family/community/committee and local elected representatives; women therefore cannot use their maximum potential and cannot fully undermine the prevailing structures of patriarchy, which obstructs their empowerment and everyday lives (Kabeer, 2011). If the number of men and women on committee were equal, women could effectively contribute and women could contribute by collectively raising their voices and male members would be forced to evaluate women's opinions.

The chairman of union *parishad* has been given ample authority in SOD (2010) to include more female members but that is not reflected in the study area. The time of the meeting and schedule for meetings is fixed by SOD (2010) and meetings are divided in different phases such as normal time once in a month, warning phase and pre-disaster period more than once a week, during disaster at least weekly, if needed daily, and in recovery phase weekly. During the FGD with the UDMC, respondents agreed that meetings were regularly held, and then while talking with UDMC members individually, meetings are irregular but they did not say so during the FGD because the chairman was present. The irregularities of UDMC meetings have been reflected in the registers of meeting provided by the Union *Parishad* Chairman's Office. In the register, the interval between meetings varies from one and a half months to five months in both Sutarkhali and Kamarkhola union.

Both the chairmen (Sutarkhali and Kamarkhola unions) stated that to ensure the presence of a majority of members, they normally fix the meeting time at a local *bazar* day which local people normally attend since prior to *bazar* time for meeting is convenient. Interestingly, the register of meetings showed that the presence of members in the meeting is extremely satisfactory and was almost 100% in some cases. The meeting documents showed that meeting regulation and minutes provided by the chairman's office that women members regularly attend almost all meetings. Sometimes, the respondents during focus group with UDMC female members could not attend meetings for reasons such as farming season, transport problems, and notices of meeting not being sent out well ahead of the meeting, household work and rainy season. Note that when we conducted FGD with UDMC researcher noticed that almost all members attended the FGD. However, the reality has been discovered in terms of the presence of women while talking with a female member after the meeting who reported the following:

> In the meeting, sometimes female members do not attend and sometimes only one or two of them attend. I attempted to put my suggestions several times but could not and the male members did not give me any attention. Currently, I do not attend meetings or even if I attend the meeting, I do not speak and the same for other female members too. (Woman, 38, class IX, FGD with UDMC)

What about signature in meeting regulations and register? Under the condition of not disclosing name, another elected women member opines the following:

> Signature does not matter; someone else is signing them off. They do not bother about a woman's signature. The signature of all members is not obligatory; rather, they want signatures from the majority who make any decision. (Woman, 45, class XII, In-depth individual interview)

This study reported that local women considered more attendance of women in the UDMC meeting, thus leading to increased participation and inclusion of women while preparing for disasters would certainly improve leadership and management capacity of women. Respondents argued that participation and involvement of women were poor in the past; however, at present, their participation in the meeting is increasing and their active involvement with different decision-making processes has been encouraged and ensured to meet different development projects for improved implementation. Note that additional participation and inclusion of women in the locality will provide better ideas and suggestions because women deal with household issues compared to men. A woman UP member describes the following:

> I guess the number of women members should be increased in local committees. Freedom and right of speech for women should be ensured. Women members are not getting sufficient space only because of the poor representation of women in the local committee, and they will not speak up. (Woman, 45, class XII, in-depth individual interview)

5.4 Voice of Women in Decision-Making Process

Respondents report that women contributed to rebuild the community after *Aila*; however, their contributions are not recognised even in the family. Women in study villages have less involvement and no involvement with the decision-making process both at family and community level, although they contributed to rebuilding the community after *Aila*. The limitations that are identified for the participation of women with decision-making process in this study are more socio-cultural and family compared to the problem with legislation and created by the society. For example, poor women are at risk because they lack status, resources and education (Fothergill, 1998; Kabeer, 2011; Morrow & Enarson, 1996). The lesser participation of women in the decision-making processes is related to family burdens, belief by the community that women are less competent, lack of funds, religious antipathy and social perceptions (Kabeer, 2011; Panday, 2016).

In this study, four reasons reported by survey respondents for ignoring the voices of women in the UDMC meeting are negligence to women (28%), importance given to males (33%), economic disparity of women (21%) and low literacy rate (19%). The problems behind participation for women empowerment in the decision-making process are recognised; these limitations require to be addressed for women's sense of self and security in the community. Moreover, it is very good that awareness is made in society for gender-neutral justice and for equality (Kabeer, 2011).

Respondents suggest a requirement to keep women in the decision-making process at both family and community level to make them confident and use their competency. The involvement of women with the decision-making process increases the capacity of individuals to transform their selection into actions and makes them able to independently work (Panday, 2017). Many respondents report that whatever projects and plans are for the coastal community, they will not be effective until the voices of women are heard and they are formed knowing women's requirements and considering their experiences. This will increase bargaining capacity, adaptation capacity, and potentiality of women in coastal areas. Kabeer (2011) argued that allowing women to be a part of projects and project plans allows changes that may be uneven because there is a greater desire for change in certain domains. For example, Panday (2016) reported that women's participation in community activities ensures equality in decision-making processes because they claim to have an improved understanding of women's requirements. Participation and sharing experiences with decision-making strengthen the capacity of society as well as family to recognise what is just and what is unjust to determine actions to be taken for a better society.

> More participation of women with UDMC will provide better management and better awareness. Better management and awareness of women will lead to better understanding about the placement of goods. Women know better where to place necessary goods e.g., food. (Woman, 39, class IX, in-depth individual interview)

Women respondents during the FGD with vulnerable women and female-headed households particularly reported that women manage households and household-related activities but cannot share and contribute for the greater interest of the community because their voices always remain unheard and ignored by the community. Ariyabandu and Nivaran (as cited in Enarson, 2000) reported the importance of the role and experience of women in their study and that flood affects existing life patterns. During a flood, while men go out of the locality, women shift cattle to safer places along with utensils from their households. They manage their children and elderly people; one woman described her experience as follows:

> My husband does not listen and does not ask for my opinion. During Aila, I kept some food in a big pot, and then placed it under soil. I did not inform my husband out of fear. After Aila, when we came back home, I dug that out and my husband became happy and that helped us a lot because there was scarcity all around at that time. (Woman, 43, Illiterate, FGD with vulnerable women)

5.5 GOs and NGOs in Empowering Women

For enabling empowerment and developing the agency of women, GO are making direct interventions via various social safety net programmes such as VGF, VGD, widow allowance and aged allowance; however, they are not as successful as NGOs. The programmes of the GO are primarily functional at a macro level. Policies and projects are initiated from the top and stakeholders from the grassroots level are not involved. The lowest unit of GO has access to Union *Parishad* level where access is inflexible and has to follow a bureaucratic system that causes failure of GO projects at a grassroots level, which has therefore been taken over by NGOs. NGOs directly work at the micro level with the basic demands of the grassroots vulnerable and marginalised people. Therefore, because of the nature of activities, women are prioritized by NGOs in different programmes. The access of NGOs to the grassroots level without following structural authority such as GOs is not authoritarian; rather, they have flexible accessibility and soft power practices. Consequently, the visibility of NGOs is higher than GOs at a grassroots level in coastal areas. The result is women have started to independently work and society at large has benefited and women have confirmed that they are quite capable of working. Micro level activities, more visibility and flexibility in power practice; priority of women, listening to stakeholders and field experience from grassroots make NGOs more successful in empowering and developing the agency of women.

5.5.1 Involvement of Women with GO Projects

Respondents report that, in the coastal area, the involvement of women in monitoring different development activities is very negligible; their presence is defined by respondents as *lok dekhano* and just by name. In response to the question "Do you think women's involvement with GO projects such as house building, distribution of cash and distribution of house building materials, vulnerable groups feed (VGF), VGD, and allowances will ensure effective implementation of the GO Projects?" almost 91% of respondents agree that the inclusion and involvement of women actively with the GOs-related activities will ensure more sustainable development and effectiveness of projects. A woman UP member described the effectiveness of involvement of local people with different projects as follows:

> We are victims and beneficiaries. We live at a grassroots level and know which initiatives are advantageous for us. People [GO officials] from Dhaka will not know the limitations. (Woman, 29, class IX, in-depth individual interview)

Moreover, women are prioritized in getting enlisted for different aid/assistance of GO such as relief, VGF card, widow allowance card, old aged allowance card and *khabikha*. Because women are the top beneficiaries of different GO programmes and projects, they should be included at the planning level. Respondents report that because GO are targeting women to be empowered, why should women not be included in the aid process rather than men? For example, women are priority on the list for relief collection but all the activities related to distributing relief are dealt with by men, including places and times for goods to be distributed. One male NGO activist described the situation as follows:

> Aid to women is to empower them and only women are listed to collect relief; women are in the priority list too. Because women are a priority to be listed, they definitely know the problems related to distribution and collecting [of aid] from union head quarters better. Therefore, they must be included because programmes are for them and they are a priority. (Male, 44, class XVI, FGD with NGO activists)

The importance of the involvement of women with GO projects is reported by a woman respondent as follows:

> GO have had problems managing local people because they come through the locality and have considerable contact with the chairman and influential people of the community and then they (GO officials) form committees for distributing relief, aid and different training programmes. Corruption primarily takes place when they include names in the list. They include the names by listening to people they came into the locality through. Both social and political identity play an important role in whether you will be included in the committee, e.g., training programmes are for women; however, the list is made from listening to a few male members. How does this lead to a better result? If they listen and involve women, the result will be better. Women know who requires a sewing machine or who needs livestock and poultry training (Woman, 65, class IX, FGD with female-headed households)

The respondents report if women's involvement increases at a large scale in the GO projects, malpractice in distributing relief, aid and works related to recovery and preparedness will be reduced; this will ensure more equity for all. Moreover, political bias will be reduced if women have active involvement in politics and GO activities. Places and strategies to implement projects will yield better results if women's opinions and experiences are taken before implementing projects.

5.6 Involvement of Women in NGOs

GOs are more responsible for ensuring governance and sustainable development. However, in Bangladesh, large-scale corruption flourishes when public officials have monopoly power over service recipients, organisations/institutions have great discretion and when the accountability of public officials to the principles is weak (Haque & Ali, 2011; TIB, 2018). NGOs in Bangladesh as micro/grassroots level empowering and community-based practice agents (Kabeer, 2011) have taken over the role of forming social capital, ensuring accountability and good governance by community empowerment involving community people and it became possible because of GO failure (Alston, 2015; Panday, 2016). However, the importance of their work lies not only in the direct aid and training they provide but also in the formulation of principles for action that guides their work and is absorbed by villagers. It became possible because of good governance by NGOs, including their sense of responsibility, ensuring rule of law, greater participation and representation in their action (Haque & Ali, 2011). This indicates the wide involvement of civil society and community people to ensure rights and standards by getting wide support by citizens for their programmes delivered for welfare at a community level (Haque & Ali, 2011; Johnston, 2017; TIB, 2018).

This study reports that local NGOs are taking initiatives for women empowerment and these initiatives are more likely oriented to cultural practices in the coastal area with accountability to local people. For example, NGOs are particularly focusing on women's empowerment and development of women's agency for sustainable development in the study area by ensuring the participation of women in developmental processes. Because women remain in boundaries of their homes and they have very limited access to means to resolve community problems, there are very few women leaders at a grassroots level. Women lack experience and skills and knowledge of political activities and national issues. In this situation, NGOs intervene to ensure women's participation in the political and decision-making process with the aim of empowering women by strengthening women's bargaining capacity. Moreover, the participatory, empowering and democratic approaches to development process by NGOs indicate they listen to the requirements of the community and they involve people from their beneficiary group to ensure accountability and transparency to strengthen capacity for good governance. NGOs have been preaching the idea of good governance such that people get a fair distribution of resources without any likelihood of deprivation induced by corruption (Kabeer, 2011; TIB, 2018). Initiatives and programmes taken by NGOs in the coastal area for empowerment and participation of women are described by an NGO activist as follows:

> Shushilan (a regional NGO) have a project titled "Flourishing of Women's Leadership." Under this project, Shushilan registers only women for membership and they come from different villages and wards. In every ward, there is a committee. The NGO provides them training particularly on gender, leadership, good governance, democracy and social issues. Shushilan activists keep them updated about national and international issues from the newspaper, television and radio, as well as arrange cultural and amusing events for ensuring their participation. However, through this process, local women are introduced to the outer world and organisations, which are working on legal rights, dowry, and early marriage (Male, 44, class XVI, FGD with NGO activists)

Another NGO activist describes the following:

> Rupantor, a local NGO has taken on "women's political empowerment in south-western Bangladesh" with a view to developing women's leadership at a grassroots level so as to have their participation and coordination with the local GO institutions (union parishad). The fundamental principles of projects are organising, educating and empowering women at a grassroots level toward a positive direction through developing leadership and capacity to contribute to the community by ensuring access to resources and justice at local level institutions and service providers and administration. (Male, 33, class XVI, NGO activist, in-depth individual interview)

By considering hazards and disasters as "regular must happen events" in the study area, NGOs targeted women as their central focus. In the study area, the participation of women with the development process is extremely low at both social and institutional level, although they have active participation at a family level to reduce vulnerability even though it is unrecognised by family and society. In the study area, 95% of respondents received aid and assistance from NGOs after *Aila*. The practical needs-oriented programmes taken by NGOs in the coastal area have been reflected on by a male participant as follows:

> We do not need loans, relief, clothes and rice; we rather require more employment opportunities. We welcome employment opportunities and earning options. Women require training on poultry, sewing, livestock, and homestead gardening. Men require training on farming such as how to use fertiliser and seeds (Male, 75, Illiterate, FGD with vulnerable community people)

The main programmes that the local people mentioned were provided by different NGOs including training and awareness (74%), relief programme (71%), loan (45%) and medical support programme (34%). The empowerment of women through earning, financial contribution in the family, and raising awareness for others have increased in the coastal area after they were getting priority for different aid, assistance and programmes taken by NGOs during a disaster, particularly after *Aila*. One of the respondents added and others agreed that:

> We always enlist really needy families and women for aid and programmes; women are our focus because we identified them as the most vulnerable group, and we think women are the best managers to reduce vulnerability. (Male, 42, class XVI, FGD with NGO activists)

The local and national NGOs undertook programmes for the advancement of women by prioritising them in different programmes, including awareness among women about rights after the declaration of the UN International Year of Women and the subsequent UN Decade of Women (1975–1985). Considering the suggestions of international NGOs and the organisations of the women's movement who act as bargaining agents for women and organise local demonstrations, NGOs bring women out of the kitchen by building agency, which started in Bangladesh in the early 1980 (Alston, 2015; Nazneen et al. 2011; Panday, 2016). NGOs distribute loans in the coastal area among women to ensure women have something that they own. Although men often take control of these loans, NGOs still give them out to women as they cannot control what is performed in the family, and they can only persuade. The participatory and empowering approaches of NGOs slowly created space for poor at a grassroots level and marginalised people in the decision-making process. This bottom-up approach and role of NGOs as "watchdog" as well as in enabling local communities turned out to be a tool for accountability for the NGOs for empowering women (Haque & Ali, 2011; TIB, 2018). The idea of NGOs giving loans only in the name of women was that this would increase women's control and benefit them by empowerment, which has been depicted by an NGO activist, as follows:

> Real empowerment and emancipation of women will not happen until she earns money and her decision will not be evaluated or valued in the family and society. From this stand NGOs targeted women in the coastal area. NGOs provide loans, relief and other programmes to women so that they can contribute equally in the family like a man. If women do not have any property and if they cannot earn, they are ignored by men in the family. But when women own money and earn money, people of the society will give value to their decision. That is why NGOs work with women. (Male, 34, class XIV, FGD with NGO activists)

5.6.1 Role of NGOs in Empowering Women

In coastal areas, NGOs showed success and improvement for strengthening and empowering women by creating earning opportunities and providing different training, including training for disaster preparedness. In this manner, NGOs reached a satisfactory position for the issues of good governance and accountability, which gained prominence in their organisations (Haque & Ali, 2011; Panday, 2016). They are extensively accepted in Bangladesh as an alternative to GO; thus, they can fill up the emptiness created by the corruption of GO. The credibility of good governance and accountability in Bangladesh has been gained by NGOs because of their macro and micro level accountability, which includes internal (accountability to common/grass root people/service recipients) and external (accountability to GO rules and funding organisations/donors) accountability (Haque & Ali, 2011; Saifullah & Ahmad, 2008; TIB, 2018).

The dynamics of women's empowerment in coastal Bangladesh have improved, particularly after *Sidr* and *Aila* because of interventions by NGOs (including training, knowledge sharing, awareness raising, involvement with decision-making processes and access to employment opportunities); moreover, they affected the livelihood and everyday practices of women at individual, family and community levels. Women empowerment and gender neutrality lead to legal and social change (Groenhout, 2002; Panday, 2016; Ripley, 2008) in the status of women in a manner that is fair, which is described by an NGO activist:

> NGOs arrange amusement and recreational events such as drama related to different social problems such as child marriage and dowry. Through recreational events, women are learning legal points of dowry and child marriage. (Male, 42, class XVI, FGD with NGO activists)

In the Bangladeshi coastal area, the knowledge about the rights and responsibilities of women provided by training and capacity-building initiatives by NGOs make them more able to face existing challenges (Panday, 2016). Moreover, this removed the images of women as "passive and obliged to be in the home". For example, after being trained, women are going to houses and talking to other women to make them aware and teach them what to do during a disaster. Women who received primary training are now helping others even if there is no disaster. The involvement of women and training for women and its effectiveness was observed in 2014 when another cyclone *Hudhud* occurred in the coastal area. Although it was not as severe for the coastal people of Bangladesh, all people in coastal areas took precautionary initiatives better than at any time before. Knowledge gained by women through training not only effected change in women's behaviour patterns but also increased their understanding of their roles and the responsibilities and rights of other women and men (Panday, 2016). Moreover, it increased the ability of members of disadvantaged communities, individually and collectively, to make decisions about strategic life choices (Kabeer, 2001a; Sen, 2000).

Many respondents reported that if NGOs play more active roles in implementing and operating their different programmes, the capacity of women will be improved, which has been reflected on different occasions, particularly after *Aila*. However, respondents reported that inclusive participation is required, particularly training for both men and women. Respondents argued if men attend training with women provided by NGOs, this will help even more to overcome societal values and existing social structure because it cannot be denied and overcome overnight. When family male members know that programmes are taken by NGOs for the betterment of community people, male members will inspire women to attend NGOs' activities and not create obstructions for women members as per the female-headed household respondents.

All respondents during FGD with vulnerable women stated that they do not face any problems in attending any programmes and meetings nowadays. The social norms, culture and practices of women in coastal areas have changed as per the respondents. Changes in socio-cultural and religious practices began after cyclone *Sidr* in the coastal area but transformed remarkably faster in the post *Aila* period. For example, as compared to the last decade and reported by the discussants, religious norms, values and practices have become very much more flexible in the study area. Previously, male family members, as well as community, were used to create obstructions for women to go out. However, UDMC members, along with male members of the family, now allow women to attend different programmes. A woman respondent explained the focus group discussion with the researcher as a reflection of religious flexibility, which has been described as follows:

> Do we know you (researcher)? We did not know you; however, we came to a meeting with you. If religious practices are strictly followed, we must not meet you. We are twelve women and no one is wearing borkha; we invited you inside our house and are talking with you in a friendly and open manner. There are no obstructions from community leaders. Moreover, UP male members as community leaders called us to talk to you. (Woman, 30, can sign only, FGD with female-headed households)

Another woman discussant explained:

> Women attend training, relief and aid programme, and food for work programme. The regular practices of going out for women for different purpose make society and local male members tolerant. It is not like there is no problem for women; however, the rate is extremely low. This is a meeting we are attending in a shrimp gher; this is an open place, which means religious practices are not rigid any more, because of the necessity of society and family. (Woman, 22, class XIV, FGD with vulnerable women)

> Priority of women, promotion of women and ensuring the involvement of women, particularly political empowerment of women at a grassroots level, certainly improved the bargaining capacity of women in the coastal area. For example, the Women Coordination Centre (WCC) is a platform to develop direct political leadership and the decision-making power from women at a grassroots level. By forming a formal platform such as WCC, three women are competing for UP chairmanship and 19 women for UP membership, respectively. Is it not a reflection of political women empowerment and is it not a reflection of the bargaining power of women in the coastal area? (Male, 33, class XVI, in-depth individual interview)

The effectiveness of women's empowerment by NGOs was reflected when the researcher met a woman UP chairman candidate who was from the local area during her campaign. For the last few years, a local NGO, namely, *Nari Bikash Kendra*

Fig. 5.3 Woman UP chairman candidate campaigning for election accompanied by two male campaigners

(Centre for Women Development), has operated different programmes such as political empowerment of women that motivate and raise awareness about women's rights. Consequently, a woman who was an active member of the *Nari Bikash Kendra* and became aware of her political rights extended direct candidature as chairman of the UP for the first time. The woman candidate, who was campaigning accompanied by a male member (Fig. 5.3), funded the campaign herself, although she gets some small support from the NGO. About her candidature, she stated:

> Pray for me please. I am fighting against powerful men. I am a chairman candidate from Nari Bikash Kendra. I want to work for women in this area. In this of region history, I am the first woman UP chairman candidate. (Woman, 43, class X, In-depth individual interview)

In its true sense, development is the outcome of rule of law, transparency and accountability, participation, contention among groups and interest in society (Johnston, 2017). Development is not merely an economic change but rather promoting and sustaining benefits produced by poverty alleviation and growth that lead to good governance and accountability (Panday, 2016). The literature suggests (Mir & Bala, 2014; O'Dwyer & Unerman, 2008; William & Taylor, 2013) that the upward accountability (pressures/monitoring by GO regulators and donors because funding by donors is subject to accountability to donors regarding implementation and return of programmes) and bottom wards accountability (accountability to the stakeholders/beneficiary because of exit and voice mechanisms) of NGOs make them more successful for empowering women and development. Rather than challenges, the participatory and empowering approaches of NGOs created space for decision-making processes at a grassroots level for poor and marginalised people. NGOs are more successful in promoting good governance, developing feelings of ownership of the programmes by community people, ensuring greater representation and participation, strengthening women's rights, independently developing capacity to work, articulating public needs and practising solutions for resolving public requirements.

5.6.2 Bottom-Up Approach of NGOs and Women's Empowerment

The drivers or agents of change can emerge or be initiated in a society either from "top-down" or "bottom-up". However, the bottom-up approach entails shifting accountability directly to citizens and service receivers by using exit mechanisms (scope of service users to exit for unsatisfactory service) and voice mechanisms (articulation of preferences and demands by the service users), argued by Haque and Ali (2011). The respondents report that projects and programmes developed by GO and NGOs for the locality must be participatory and local experience-based (bottom-up) because women who have survived disaster can contribute better in disaster management and disaster resilience process; after all, projects are for local people. However, the problem for Bangladesh is that the top leaders/decision-makers (GO) are not committed to preventing corruption or indeed are indulgent towards corruption, naturally, the lower rungs of leadership indulge in it (Islam, 2015; Kabeer, 2011). This study shows that pressure and expectations from the bottom for implementing the programmes make NGO programmes effective, which is reflected in Kabeer (2011) and Panday (2016) as the successes of NGOs' stem from the pressure of ordinary people. A male respondent described the effectiveness of a bottom-up approach as follows:

> Bottom-up approach must be effective for disaster-prone areas in the coastal community and vulnerable people in coastal areas because people at a grassroots level are much more aware than in the past. Because of the lack of participatory discussion and sharing knowledge with local people, many projects and programmes are being ended without results. (Male, 45, class XVI, FGD with PIC)

Messer (2003) argued bringing GO's decision-makers closer to citizenry is extensively believed to increase public sector accountability and effectiveness while contributing to the strengthening of a genuinely people-centred democratic culture.

> The effectiveness of bottom-up initiatives will be extremely fruitful because the plan, design, demands, and process of implementation is attributed to people at a grassroots level. For example, to develop a project for a particular village in the coastal area, we require a budget. Therefore, to determine the budget, we need to go to the village and talk to the people at a grasroots level to know what they want to do for the village. Then, we will add components as per their needs and priority of needs. (Male, 48, class XIV, FGD with NGO)

Ribbot (1999) argued that participation means bringing a broader section of a given population into the public decision-making processes, which is particularly important in the context of mitigating disaster-related risk. Greater participation in public decision-making processes and community interest lead to improved positive results such as efficiency and equity in developing resources. In other words, experience and participation of local people with the development process are incorporated to ensure effective development and gender-neutral disaster management strategy. For example, the bottom-up approach will be more effective for disaster management and development if early evacuation training is arranged for women and local people and if local rescue teams are formed by local residents. However, the importance of a bottom-up approach is illustrated by a male discussant during a focus group with project implementation committee members:

> To ensure projects and programmes are successful and for the best output, support of the local people is necessary. It is much easier to continue projects and programmes in a fruitful manner with the collaboration of men and women of the locality. (Male, 27, class IX, FGD with PIC member)

Crook and Manor (1998) argue that bringing GO closer to people increases efficiency by helping to "…tap the creativity and resources of local communities…" In summary, bottom-up means participation, coordination, experience-sharing and access to local skills and information using local knowledge in every step of society. Many grassroots level initiatives of women demonstrated the value addition that women make for reducing the disaster-related risks through preparedness measures and for rebuilding communities affected by a disaster.

This study reports that a positive association among community, women and GOs would make all projects sustainable and bring more benefit the people in coastal areas. When asked "what makes women feel that the involvement of local women with GO activities will make GO activities sustainable?" It is being identified by the respondents that easy access of women (57%) to the beneficiary/households of the GO projects provides insights into problems and better understanding of women's problems by other women (19%), which will in turn make GO activities sustainable and effective. This shows that the involvement of women with earning (11%), less involvement with politics and political activism (11%) and knowledge from good household managers (10%) make GO activities successful, sustainable and effective. Panday (2016) reported that women established themselves as "less corrupt", "more committed to serve the people", "understanding the needs of other women better" and "politically less biased" than males (Fig. 5.4).

5.7 Challenges in Empowering Women

GO and NGOs play an important role for socio-culturally grounding women's empowerment in the coastal area. For this purpose, they prioritised the marginalised group such as women at grassroots level by including them in different decision-making platforms. Despite positive impact in empowering women both at macro and micro level taken by GO and NGO, there are challenges that limit the speed of the whole process of development, although the nature of challenges for GO and NGO differ. For example, for implementing their programmes, GO are primarily obstructed by their structural hard bureaucratic power system, whereas NGOs have a soft and flexible authority system with access to individuals at a grassroots level. For successfully implementing projects and for empowering women, the challenges for GO are identified as the ignorance of local people's opinion, corruption, misconduct and bias, which are created by GO officials themselves. However, the challenges for NGOs are identified as a lack of coordination and cooperation of GO with NGO, in

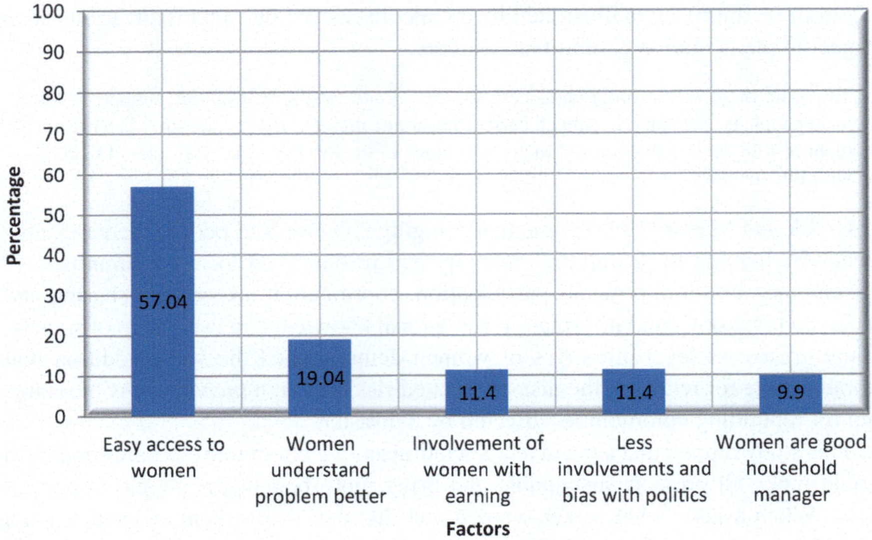

Fig. 5.4 Causes will make GOs projects more effective for women

addition socio-cultural and infrastructural factors. This section compares challenges for both GO and NGOs for empowering women in coastal areas.

5.8 Challenges of GO

5.8.1 Ignoring Local People by GO in Social Safety Projects

The different social safety net programmes by GO have lifted normal standards of living of millions of people in coastal areas during the disaster recovery phase in an increasingly effective manner (Alston, 2015), which is reflected in this study area where 100% of respondents report that at disastrous times they are extremely dependent on GO aid projects, although GO projects are not as effective and successful as they are supposed to be. The respondents report that GO officials do not know and do not consider the coastal societal system, practices, experiences and norms at the time of projects' planning and implementation. Survey respondents report that GO officials do not have local area knowledge (75%) and not take the opinion of local people (70%) before making plans and implementing a project. Consequently, GO projects cannot achieve their goals and become unsuccessful. In this regard, Alston (2015) argued that strong governance structures at a local level are important to effectively aid and distribute resources. While conducting the focus group with members of PIC who are supposed to be actively involved with project planning and programmes, one male member reported the following:

> We do not know details about programmes taken for our village. They [GO officials] do not come in our villages before implementing any projects. If we assume they are here to talk to us, we are sharing our views with you. If the decisions are obtained in this manner, projects and programmes will be more effective for us. (Male, 48, class XVI, FGD with PIC members)

The respondents report that GO officials did not see *Aila* and did not face flood and cyclone at the actual time, they just guessed based on the reports of media and different NGOs. However, based on their assumptions, they start designing and planning projects and subsequently implement projects. A male respondent reported the following:

> Normally, GO officials visit the affected area for a day or for couple of hours during a disaster. Whenever they visit the affected area, common people of villages cannot go to them and GO officials do not come to speak with the general people. While visiting, GO officials are surrounded by local influential people, including political leaders, UP chairman, and members and wealthy people in the community. (Male, 37, class XII, FGD with PIC member)

Another male member of PIC described the following:

> Indeed, we are involved but that involvement is after project planning and does not start before planning of the project. GO had taken a project to develop houses in Sutarkhali for people who completely lost their houses during Aila. We provided our opinions while GO officials came to identify sites to develop houses but our opinions were not considered when executing decisions. For example, we asked them to build houses at a good height. We showed them the height level from the experience of water level from Aila to avoid flooding the house in future because of cyclones but nothing has been executed as per our suggestions; rather, they completed projects as per plan they had made. (Male, 50, class XIV, FGD with PIC members)

The respondents report, for example, last time after *Aila* GO officials visited the community for a couple of hours to determine the site on which to build houses and consequently the site was selected with the help of the local leaders and influencers. Sharing meetings and exchanges of views between GO officials and local people about house building projects must be more effective and sustainable for the people who live in them. Sustainable development in the true sense is the outcome of participation, contention among groups and interest in society (Johnston, 2017). Alston (2015) described that stable, fair and representative decentralisation of disaster responses to the local levels of organisation can draw on local knowledge and assist in developing the ongoing mitigation efforts. A well-organised local level of governance, even if inactive during stable periods, can ensure comprehensive attention to reconstruction during a crisis, which has been reflected in the survey results of the study. The majority of survey respondents (86%) reported that local people must be included and their experiences should be strongly considered for accountability to ensure more sustainability of GO projects.

During FGD, a respondent was saying that they are helpless and powerless in the sense that they are getting houses and other aid but not as per their suggestion and demands/requirements. For example, houses built for people who lost their houses after the last disaster have been rebuilt on low-lying land without pillars. If houses had been built on pillars, water could easily pass by and they would be more resilient. Unfortunately, this did not happen as GO officials did not consider the experience of local people before making and implementing the plan. Furthermore, this reflected their powerlessness because local people did not have an opportunity to share their experiences.

Panday (2016) reported that development is not merely an economic change rather promoting and sustaining benefits ensuring greater participation, which leads to good governance. The ignorance of knowledge and voice of local people by GO officials created a gap between local people and GO officials; this is reflected in the following statement of a male member of PIC "*it is Bangladesh and it is apparent that there is a gap and this gap is created by the GO officials and it is not to be solved*". This argument shows that the ignorance of local people's opinion and experience and exclusion of them by GO officials from the projects that are created. Exclusion is created to show off that they are superior and powerful. They are inaccessible by anyone at any time; therefore, they just come to visit the affected area and community for a day and sometimes for a few hours. There are positive aspects to this realisation by local people and identifying reasons for their exclusion shows the agency of the local people; however, one of the respondents described his experiences as follows:

> Have a look on the roof of my house; it is made with cement, concrete and rodsin Dhaka and then brought in the area from Dhaka. This village is 500/600 km away from Dhaka. If they would make this in the local area, it must cost less and the local needy people could work as labourers. We could not remove the roof when we had a problem with it because it is fixed and permanent; it is extremely heavy to remove. It could be handy, flexible, and removable if it were made of tinsheet as well as if it had been composed of sunsheet we could repair it. (Male, 55, class VI, FGD with PIC members)

Discussion with respondents demonstrated that if there had been a collaboration meeting between GO officials and local people; if local people knew about GO's projects, the local community could even have been financially benefited. For example, houses distributed among homeless people in the coastal area were built in Dhaka and transported all the way to the study area; however, if they were built in the local area it would create an opportunity for jobs and it could also reduce transport costs (Fig. 5.5).

5.8.2 Corruption, Misconduct and Bias of GO Officials

Corruption by GO indicates the deviation of GO officials and institutions from certain behavioural standards from their public role and public interest, which includes bribery, nepotism, deception, negligence, harassment and misappropriation of assets with a deliberate intention to abuse public power and interests for private and personal gain (Iftekharuzzaman, 2011; Knox, 2009; TIB, 2015). Survey respondents report that nepotism, politicisation, personal relation and corruption by GO officials are manifested in different projects such as financial support (83%), food support

Fig. 5.5 Distribution of perceived feelings in terms of involvement of local people with GO project and sustainability of GO projects

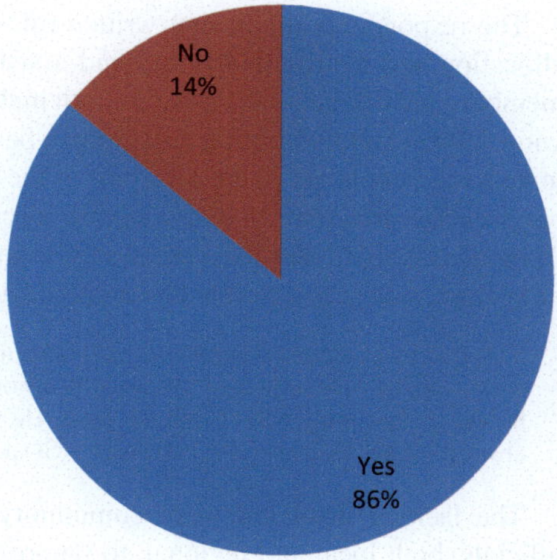

(69%), *kabikha* (19%), distribution of construction materials (10%), medical facilities (11%), clothes (11%) and agricultural accessories (4%). The following remark reflects the practice of bias and discrimination:

> The distribution and allocation of GO are discriminatory. After Aila, I received only 20kg of rice from GO; however, some other neighbours received extra benefits such as cash support. They got the list ready after staying at their office and knowledge from a few local people, including the UP chairman. They distributed among people as they wished. Physically fit people became disabled; moreover, disabled and sick people became fit. (Male, 60, class X, FGD with vulnerable community people)

The respondents report that GO allocated 20,000 taka for each family of the locality. Cash was distributed by the elected local union *parishad* members who kept certain money from the allocated 20,000 taka but no one raised their voice in case later on he or she did not get any aid at all by way of retaliation. Moreover, there was a GO house building loan project to develop and rebuild houses. However, only people with lands for cultivation were eligible to get the loan, which is discriminatory. While discussing with UDMC members, the discrimination and irregularities in distributing aid and assistance, all members disagreed and raised the following counter questions:

> The relief card is in the name of women and women are in a priority list of different projects such as Kajer Binimoye Khaddo and Mati Kata. How do you expect discrimination for distributing relief? There was no such discrimination; rather, women got preferences (Male, 47, class X, FGD with UDMC)

In this argument of the UP member, while conducting the focus group with vulnerable community people later on a male discussant said:

> To be involved with GO earning activities such as Kajer Binimoye Khaddo and Mati Kata, relief aid card and vulnerable group feed card, women have to go to UP chairman and UP members. To get this card, women have to give a bribe ranging from 500 to 1000 taka. However, no one will acknowledge and confess it because they are scared of losing the card. (Male, 37, class VIII, FGD with vulnerable community people)

The respondents report that written rules, systems and directions regarding the allocation and distribution of aid and assistance are not public; consequently, UP members and officials can easily commit malpractice. Moreover, local people neither know how aid and allocation are going to be distributed nor how much they will get; in general, people are not at all a part of the distribution process. Ignorance of local people about allocation and distribution process and keeping the general public away from the distribution process creates an opportunity for people with responsibility to be corrupt, which is described by an NGO activist as follows:

> GO provides all aid and programmes of the affected area to the UP chairman and the chairman deals [with it] with the help of members. If any one's wife/husband is close to a UP member, he/she or the family will benefit the most. He who gets aid, gets it from all sources and always. (Male, 48, class XIV, FGD with NGO activists)

The lack of involvement of community people in GO projects provokes GO officials, both local and regional, to ignore transparency and accountability, which promotes bias, corruption and irregular practices. However, survey respondents identified different types of bias, including nepotism (83%), politicisation (51%), cheating in weight (18%), personal relations (52%) and women kept unnecessarily awaiting (11%). Poor governance and corruption are the primary reasons for the gap between citizens and service providers in other studies (Islam, 2015, TIB, 2015). Relief does not reach the poorest people who really require it badly and the requirements of women are generally overlooked because the poorest group and women are economically and politically disadvantaged, indicating that there are different levels of poverty even among the poor. However, this study shows that women are the most vulnerable in coastal areas during a disastrous despite having aid and support from GO and NGOs. However, bias by GO officials and UP members makes them more vulnerable because corruption, patronage networks and misallocation slow down the economy and prevent developmental leap (Bertelsmann Foundation, 2012; Islam, 2015; TIB, 2012; 2015), which is described by a vulnerable woman as follows:

> I got a house from the GO but it is not everything. I do not have sufficient earning to bear other expenses and cost of my family. I cannot meet the basic demand of my family members such as food and clothes. I stopped sending my young children to school. I am not getting aid and benefit of the GO allocated for needy women in the coastal area. Whether it is possible [for me] to sleep peacefully does not matter [to the government], I got a house. (Woman, 29, class VIII, FGD with vulnerable women)

When asked during a field survey what challenges do women face, the challenges during a disaster include threat by local leaders and influential persons of society (18%); GO officials and local political leaders asked the poorest group and women who are economically and politically disadvantaged for a bribe (21%). GO officials have always been high up on the list of obstacles for doing business, which encourages use of bribes to speed up administrative processes (Knox, 2009; Wickberg, 2012; Bertelsmann Foundation, 2012; TIB, 2018). Respondents argue that corruption as a social process, such as bribery, is established and in areas such as public resources is maintained by a selected group of society, usually described as elite.

When researcher was discussing challenges with the UDMC in the focus group, one of the vulnerable woman discussants described that bribery and threats by local leaders and GO officials took place most commonly at the time when houses were being built on the embankments/dams or in any other abandoned places of GO. Some other women discussants during focus group with the vulnerable women and female-headed households responded that sexual harassment is a challenge (16%) for women during a disaster. They argue that women will not disclose their challenges but they are common, e.g., when a woman goes to the chairman and members of UP for permission to put a house on GO's abandoned land sometimes they do not give any decision in front of other people and instead keep woman waiting for nothing and allow other people to go. A vulnerable woman talked about an illegal offer as follows:

> I felt upset while I sought help from a UP member after Aila. One day, I met this member and I was alone. He touched me and I came back without any aid. I could do illegitimate things for aid but I controlled myself. I have a daughter and I want to arrange her marriage with a good family. (Woman, 27, class V, FGD with vulnerable women)

Poor people are extremely vulnerable to corruption by GO institutions and are penalised most because of their dependence on services (Islam, 2015; Kabeer, 2011; TIB, 2012, 2015; Wickberg, 2012). Out of the total number of respondents, 41% said that money is a challenge while 22% identified bias and nepotism along with scarcity of food and lack of cash money as challenges during a disastrous in the study area and only 6% of respondents replied they did not face challenges. One vulnerable woman respondent after a focus group discussion informally raised an important issue regarding challenges during a disastrous as follows:

> Challenges after Aila were common but challenges became unequal and disproportionate while rehabilitation was under way. Relief, aid, and different assistances were not equally distributed and certain people were more privileged and this happened because of certain corrupted people of the locality who were biased. (Woman, 48, illiterate, FGD with vulnerable women)

The malpractice and bias by GO officials and UP members have been identified by one UP member of Sutarkhali as a "favour for some people" who are from the local community and "favours" have been described as follows:

> Listen brother, you are an educated man and you know everything. We are elected and certain people worked hard for us before an election. They looked after us and now we look after them. We have a political party and party boys' expect something from us and it is not discrimination. (Male, 45, class X, FGD with UDMC)

In other words, "favour" of the UP member is basically a biased exchange relationship involving a largely purposive/need-based friendship in which an individual of higher socio-economic status (UP members in this study) uses his own influence and resources to provide benefits for a person of lower status who, for his part, reciprocates by offering general support and assistance, including personal or political services (Chowdhury, 2015). In Bangladesh, the relative burden of corruption in the name of bias and favouritism is high for the poor. They bear the additional burden of corruption than the people with higher income; it disproportionately hurts the poor too (Iftekharuzzaman, 2011). This study shows that favour/favouritism, bias, corruption, discrimination and misapplication of resources by office holders disrupt community collectivity, fuel up community conflicts and undermine social cohesion, which is reflected in other studies (Saha, 2015). How does favouritism impact social cohesion? In response, while conducting the focus group with vulnerable women, they reported that it is apparent that there is discrimination and favouritism for certain people when distributing different forms of aid. People who are getting extra benefit and favour from people involved with distribution are not positively accepted by other people of their society who differently treats them, which ultimately impacts their neighbourhood relations. When abuse of power is not punished, individuals and members of political elite in positions of power do not follow the values of common good of society; if they determine the limit of their behaviour in Bangladesh, they are then responsible for corruption. However, in the rural area of Bangladesh, the accountability for financial and other support is limited by power exercised by the local elite, elected representatives and political leaders (Haque & Ali, 2011).

5.9 Challenges of NGOs

5.9.1 Lack of Coordination and Cooperation by GO to NGOs

Despite good understanding and familiarity of NGO activists with social structure and societal values, the programmes and initiatives taken by NGOs are not as effective as they could be. Respondents reported that the success of NGOs in the coastal area could be even better if they could get all out support and coordination from the GO. A respondent reported that the gap between supply and demand of community would be solved better if there was coordination between GO and NGO. An NGO activist described the following:

> We (NGOs) are partner of the GO for implementing this project. As partners, NGOs finish projects with additional responsibility and coordination. For example, all workers, including me, as a supervisor came to the site 20 minutes prior to scheduled time to start work at 7AM. Someone else is monitoring the activities in the field, which is extremely rare for GO projects. (Male, 25, class XIV, in-depth individual interview)

Because of the nature of work and involvement of NGOs with community people, they are trusted more by community people. Consequently, NGOs reached a satisfactory point for issues of empowering women and good governance and gained prominence (Haque & Ali, 2011; Panday, 2016). NGOs are widely accepted in Bangladesh as an alternative to GO that can fill up the emptiness of corruption (Haque & Ali, 2011; Saifullah & Ahmad, 2008; TIB, 2018). This study shows that the effectiveness of NGO activities and initiatives depends on their capacity to rapidly reach community people at a grassroots level with aid and assistance, and the success lies in familiarity with social structure and societal values and familiarity with the concerned target group. NGOs in Bangladesh primarily operate their programmes through direct engagement at a grassroots level for women, poor and marginalised groups (Panday, 2016). This helps NGOs be oriented with societal values. In terms of familiarity with societal values and social structure, a male respondent reported the following:

> NGOs deal with people at a grassroots level and an NGO activist works with rural people regularly because projects of NGOs are based in rural areas. They know each other better and have improved understanding. Because of familiarity with social setting and coastal people, NGOs are more effective for coastal people to empower women and reduce vulnerability. (Male, 34, class XIV, FGD with NGO activists)

Although NGOs gained credibility of good governance and accountability, GO very often do not coordinate with NGOs because there is a high level of distrust between GO and NGO sectors in Bangladesh. For example, GO could consider the experience and share the experience of NGOs when they make plans for implementing any projects and when they come to visit the village to verify what the villagers require. GO has the attitude that they are a regulatory body primarily concerned with ensuring that NGO activities are lawful and do not conflict with GO activities and policies (Mir & Bala, 2014; World Bank, 2002). If GO and NGO develop a partnership considering the experiences of NGOs, the respondents report that programmes of empowering women certainly will be more effective.

5.9.2 Socio-Cultural and Infrastructural Challenges

What are the problems and challenges for NGOs to strengthen and increase the number of their programmes for empowering women? In response to this question, the respondents of this study identified problems and challenges, which are classified as infrastructural challenges as well as social, structural or societal challenges. Infrastructural problems include transport and communication (32%) for NGOs to reach the local community and lack of suitable training places (26%) for community people in the study area. The lack of suitable and permanent training places and backward transportation prevent women from coastal areas from attending NGO programmes and training. However, social, structural and societal problems include *purdha* system (7%), obstruction from family members (10%), and time limitations of women (34%). Because women are responsible for household activities, they cannot attend NGO programmes for a long time. In terms of these problems, a male respondent reported the following:

> If there is a fixed and permanent training place in the locality, schedules for training and meeting with the community people could be fixed and would help women to manage household responsibilities before the meeting and training day and women could attend training programmes too. Furthermore, transport and communication to attend the meeting is a problem. Walking is the most common approach to attend the meeting. Sometimes, women are require to walk 1 h to attend the meeting and training; if it is raining, women can

> not walk and it gets difficult to attend a meeting. (Male, 32, class VIII, FGD with vulnerable community people)

> After Aila, a local NGO wanted to arrange a meeting with women in the village named Joynagar; however, a local honourable person did not allow the NGO to arrange the meeting with women and insulted the NGO officials. Then, the NGO officials went back and did not distribute aid and relief in that particular village. However, when women reported that they were not getting aid and relief, they came out and protested. Consequently, NGOs were welcome for meeting with women and the women received aid. (Male, 52, class IX, FGD with UDMC)

In Bangladesh, men traditionally assert dominance over women; it results in the practice of hegemonic masculinities that obstruct gender equality. However, the UN declaration, GO policies, demand of donors and international NGOs, the voice of civil society and the women's movement in Bangladesh assisted as milestones for NGOs to recognise women and achieve equality. By addressing the local requirements, NGOs play a pivotal role in making people realise the importance of women's participation for all development processes. Efforts by NGOs open up the avenues for women's participation in the coastal area and they are becoming conscious on different issues. For example, groups of women began to demand autonomy and opportunity for social change and fight for their equality (Puigvert, 2003), which is reflected by an NGO activist, as follows:

> In the upazila health complex and hospital, there was no doctor and pathology lab for women. Initially, women in the community notified the crisis to UNO but that did not work. A couple of months later, women brought out a procession and made a human chain and met UNO and then immediately demanded a doctor for women and ensure other facilities. (Male, 34, class XIV, FGD with NGO activists).

This transformation that occurred in women's lives at a community level in the context of their status, role, knowledge and access to resources, makes them increasingly more capable of transforming their environment and articulating demands (Fig. 5.6).

5.10 Summary of the Chapter and DECR Model

In this chapter, the above discussion addresses the belief that the roles of women are different from those of men. The undermining of the status of women is not fixed by cultural norms; rather, it is shifting in the process of development via construction and reconstruction of socio-cultural practices in everyday life. Previously, the degree of participation of women at community level, public forums, UP committees, decision-making meetings, training programmes and outdoor earning activities was less. It was disproportionately affected by existing socio-cultural norms and practices. However, in the coastal area, women have recently become more capable of defining the necessity of their acts to contribute to the community betterment, which is a reflection of their empowerment. Moreover, women have gained a position to make selections on their own and transform their choices into real actions via

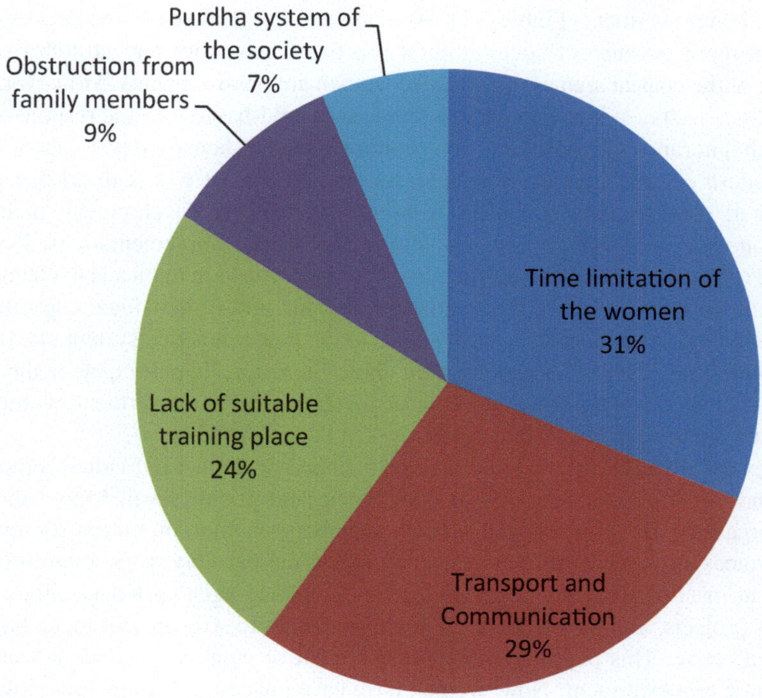

Fig. 5.6 Problems faced by NGOs to train women

participation in politics, attending meetings, raising issues in meetings, participating in decision-making processes and becoming involved with GO and NGO projects. Note that additional participation of women in public forums and direct involvement with GO and NGO projects certainly empowered and developed the agency of women.

Women are politically less partisan and involved less with political activism and being less involved with social activities such as social clubbing compared to men in the coastal area. The lack of involvement of women with partisan politics, political activism, social activities and their position as less privileged/marginalised minimise the chances for women to be biased to any particular one or any group. In addition to less involvement, the household management experiences both in disastrous and normal times and makes women accountable more than men, which positively works for good governance in implementing different programmes.

Women empowerment is revealed as a multi-dimensional process in the context of different socio-cultural practices in coastal Bangladesh. In this study, women empowerment was a part of continuous development. In the last few decades, the roles and status of women have improved because of different programmes led by different GO and NGOs in the study area, including psychological, economic and political empowerments (Table 5.1). Women were not involved in the development process more because of socio-cultural practices, behaviour and attitudes of the people of the coastal area as they believe women are passive agents. Moreover, they are obliged to stay at home and have a fixed household-based role and responsibility.

Such ignorance of women and contribution to comprehensive development in the Bangladesh coastal area has been addressed; moreover, it was realised that problems will remain unsolved unless women participate in development, including decision-making, earning, project planning and project implementations. Psychological (changes in mind-set, attitude and beliefs) empowerment leads to changes in the status of men's minds. They realized that women have individual capacity and organisational competencies to gain access to the power practice system at a family and community level and transform their choice in action. To participate in the realisation of powerlessness of women and for psychological empowerment, women are currently a part of decision-making process in the coastal area.

They are already working as committee members with their individual competencies and organisational capacity at both public and private level. Moreover, they are bargaining for position and women members on the committee for making their voices stronger. Furthermore, women have the capacity to raise questions in different meetings and ask about projects and meeting agendas because they are a part of projects, and they believe the inclusion of more women will make projects more effective. This psychological empowerment of women speeded up women's economic empowerment. Now, women have easier access to earning activities and to local *bazar* that enhances the ability of women for performing community work. They are looking for economic opportunity and creating opportunities on a small scale because additional economic independence makes them more confident. As part of the economic empowerment, women are now being actively considered as a part of social safety projects implemented by GO and NGOs. They are members of monitoring and project implementation teams. The economic empowerment has resulted in additional efficiency of women which nurtures their desire of the movement, political freedom of the women, and networking by forming formal and informal organisations. Women empowerment through participation in different platforms such as UDMC, GO and NGO programmes ensures good governance; it has sped up the political empowerment of women in coastal Bangladesh. Competing directly in elections, campaigning in elections accompanied by men, and demonstrating for their rights and justice in the study area reflects the political empowerment of women.

Women empowerment by changing different aspects of women compared to the recent past changed the condition of their lives and transformed their choices into outcomes because of intervention by GO and NGOs. They considered that women are agents of change. However, for empowering women and developing agency, NGOs have been more successful than GO despite challenges, including transport and communication problems, suitable training places, the *purdha* system, obstruction from family members and time limitations of women. NGOs are more effective than GO for empowering and building the agency of women as well as minimising gap because of their position, proximity to the grassroots people and explicit commitment to a social justice agenda. The friendly attitude of NGOs to stakeholders against the monopoly authority of GO places NGOs in a better position for empowering and agency building of women.

Furthermore, as a micro-empowering agent, the credibility of good governance of NGOs is more successful than GO because of their internal accountability, accountability to the stakeholders and external accountability, and accountability to donors and GO. However, exit mechanisms, freedom of stakeholders to quit and to complain to higher authority because of easy-going access if they are not happy with service and voice mechanisms, opportunity to raise voice by beneficiaries make NGOs successful and effective in empowering and building the agency of women. By nature, NGOs are involved in providing community services and operate their programmes by applying bottom-up processes to implement and facilitate projects. They prioritise the knowledge/experience of community people before implementing any projects for effectively building women's capacity. Empowerment and agency of women enhance the ability of women to play an important role for social change. Similar success has not been achieved in terms of good governance and accountability for implementing projects by GO service providers because of their top-down approach along with their bias, nepotism, partisan involvement, lack of accountability, superior authority attitude, absence of competent local GO staff and ignoring knowledge/experience of the beneficiary.

Nevertheless, women in the study area are involved in the decision-making process at both family and community level. They attend different committee meetings, including UDMC, PIC and CPP, although their number is poor compared to the total membership. For example, attending the meeting of UDMC, women contribute to aid distribution among women beneficiaries more transparently and effectively by sharing their experience and knowledge. While attending PIC committee meetings, they share their experience of how houses are built, what should be the structure of houses and the quality of materials from their experience of house management. Attending different meetings and knowing different issues such as effectiveness of disaster management, women easily raise awareness among other women because of their easy access and easier ability to deal with certain events such as pregnancy. Moreover, women from female-headed households, women who are undergoing economic hardship, as well as UDMC, PIC, and CPP members participate in income-related activities outside the house. In the study area, women move outside their house on their own for shopping for family, selling goods in the local *bazar* and running small businesses that are in walking distance and where all people are from the same community or known to each other. Being directly involved with NGO activities, committee meetings and decision-making processes, women participate in disaster management and disaster management policies in the study area.

In coastal areas, both GO and NGOs have extended their training and services facilities, food for work, creating employment opportunities and involving women with earning activities and infrastructural development. Activism and enforcement by GO and NGOs brought changes for women in fundamental challenges such as ideological emancipation/flexibility of the male and society towards women in the coastal area. It is easier for NGOs because they work for grassroots people throughout the year.

Women empowerment is a collective/holistic issue that is closely tied with socio-cultural practices and societal traditions. In this study, women empowerment is viewed from a holistic perception considering the involvement of women in the past few years with different aspects (Table 5.1) in the coastal community. Women empowerment was starting to achieve ongoing social transformation in the coastal area that developed the agency and bargaining power of women. However, empowerment does not mean gaining freedom in women's lives; it rather includes the ability of women to independently raise their voices against challenges, oppression, repression, exploitation and subordination. Furthermore, it includes realisation of one's strength and capacity to bring changes in the existing socio-cultural systems and fixing strategies to face challenges.

For example, it includes the requirements and concerns of women; moreover, it involves the women with various projects and programmes for sustainable community development. Moreover, this study reveals that women in the coastal area play roles in structural, social, institutional and economic development, in addition to policies and adaptation processes. This social transformation by GO and NGOs contributes to emancipating coastal community individuals, particularly women, from their discriminatory status. Furthermore, it has developed an alternative discourse of development, ensuring the presence of women in different decision-making platforms such as UDMC, PIC and CPP. Policy makers realise that without being closer to the beneficiary and broader section of the people, social transformation will be impossible. In the decision-making process, greater grassroots participation speeds up sustainable development. As a process of change, policy-makers address the opinion, suggestions and experience of community people by bringing them in the mainstream of development processes and other programmes. Effective disaster management by involving only men is impossible because they are not the only agent and they are not quite capable of fulfilling promises to empower communities and transform lives of individuals.

Table 5.1 Dimensions of empowerment and change in the roles (DECR) of women in coastal Bangladesh

Dimensions	Example
Socio-cultural	Flexibility in religious practices, changes of male attitudes, education for all, free education for girls, women's freedom of mobility, women's access to public and social spaces, participation in extra familial groups and programmes, involvement with formal and informal social networks and participation in domestic decision-making process
Economic	Food for work programme for women, earning opportunity by doing poultry, livestock, access to and control of income and family resources, relative contribution to the family, women's access to employment opportunity and access to market
Politics and policy	Access to political system, representation on the committees, public forums, promoting women's leadership, representation in local GO and access to decision-making platforms
Services/facilities	Free education for girls, relief and aid in the name of women, attending training provided by GO and NGOs, permanent sources of drinking water in the community, providing rain water harvesting technology, knowledge of legal rights and legal support by GO-NGO against violence
Psycho-social	Self-esteem after getting training and earning, well-being as part of decision-making process, collective awareness of the community to allow women to be part of development, flexibility in *purdha*, community mobilisation for women's empowerment and campaigning for rights and awareness

Source Created by the author based on research results

Chapter 6
Transformation in Women's Role in Reducing Disaster Risk and Vulnerability

6.1 Introduction

Speaking at the Rio+20 Sustainability Summit in 2012, the head of UN Women, Bachelet (2012) reported that inclusive economic growth is necessary to address climate change, rising inequality, economic uncertainty, environmental protection and social equality. This requires women's full participation, and this study reports that for various reasons, including economic restructuring, economic development, financial crisis and growth of export industries, women are now more likely than any previous generations to work outside the home. Moreover, this challenges conventional customs and raises questions about honour and male control.

This study shows that in Bangladesh, the status of women is significantly increasing because of the improvement of women's access to skill development training, job opportunities and resources that truly reflects the gradual securing of gender justice (Alston, 2015; Panday, 2016; Reddy, 2016). Women in the study area do not want to be kept from entitlements to resources, ignored in policy planning and excluded from decision-making. However, in recent decades, advances in women's status have occurred in Bangladesh and women consequently want the dignity of their lives to be respected and to have the capacity to make their own life choices, to marry when they are ready to marry and to undertake livelihood strategies using their own resources. They want to have freedom of movement, to have the capacity to protect their children and to nourish and nurture their families and communities. There is now near parity in primary education access for girls and boys and women occupy 20% of parliamentary positions (Alston, 2015). These changes became possible because of material and non-material cultural changes in society and the individuals' ability to define aims, act upon them and make decisions, which has been defined as agency development (Panday, 2016).

The evidence of changes in everyday practices in the study area and decrease in vulnerability of women in disaster times, particularly after cyclone *Aila*, is clear. Women are moving beyond the home, drawn by the necessity to earn, doing business

M. M. Khan, *Disaster and Gender in Coastal Bangladesh*,
https://doi.org/10.1007/978-981-19-3284-7_6

in the market, confidently raising their voices at UP meetings and creating additional income-generating activities. When women have their own income, a degree of autonomy and freedom enters other aspects of their lives (Alston, 2015; Panday, 2016). Respondents report that vulnerability of women, gender relations, social norms and practices improved alongside other changes because of acute scarcity of livelihood options after massive scale destruction caused by *Aila*. The scarcity of livelihood options and scarcity of necessary goods forced men to change their attitudes to women and primarily allowed women to get involved with earning activities and to work outside.

More women, including those who are heads of their households in the coastal area, are involved with income-generating activities and these major changes and pace of change have been created by a climatic challenge. Climatic challenges have created hardship; however, this has led to a critical requirement to provide a more empowering space for women and respondents reported that, in the coastal area, cyclone-related death, damages and loss have been reduced. Increasing awareness at a grassroots level on a large scale compared to the past by both GO and NGOs, particularly by NGOs, has been successful by introducing a "people-centred approach" to the GO policy and beyond (O'Donnell & Wodon, 2015).

GO-NGOs as partners have taken multiple programmes, including local-level planning, asset creation and employment generation, capacity-building of local stakeholders, asset grants for productive investment, disaster risk reduction and life skills training, with the aim of building resilience of households and communities, improving food security and improving economic resilience where women were being prioritised (GoB, 2012). Moreover, NGOs with the support and coordination of local GO have undertaken projects for promoting gender equality via women's political empowerment and developing grassroots women's leadership at union *parishad* level (Rupantor, 2013). NGOs and women's organisations are central in making an effort to change ordinary people's attitudes towards gender roles (Panday, 2016). The respondents themselves argue that natural forces such as climatic variability have changed the existing established and long-practised gender relations.

6.2 Changes in Maintaining *Purdha*

Baden et al. (1994) report that, in the 1991 cyclone in Bangladesh, out of the flood-affected population in the 20–44 age group, 71 females per thousand died compared to 15 males per thousand. Kabeer (1990) reported the argument that *purdha* creates a sexually segregated world that identifies men with the public/social sphere and women with the private/domestic sphere that result in the vulnerability of women. *Purdha* is grounded in structural constraints that deny women access to both social power and any form of autonomy over their own lives (Kabeer, 1990). However, vulnerability to socio-cultural inequality, including *purdha*; dependency on men's wages; unequal access and an unequal share in husbands' incomes started to slowly improve from the mid-1990s. Moreover, after the mid-1990s, women in Bangladesh have started to exemplify their success of economic empowerment at an individual level by going out of the house. Women experienced mainstreaming from collective aspects that addressed women's political emancipation, economic empowerment and advancement of status. Women empowerment was emphasised when NGOs, donors and activists focused on women as an inclusive part of society; they conceptualised that changes for women are changes for society (Nazneen & Sultan, 2012; Nazneen et al., 2011; Panday, 2016). Mainstreaming and inclusion of women and changes in socio-cultural practices have improved women's ability to make choices, which were denied in the past. Moreover, they gained ability to control multiple aspects of their life, including defining aims and acting upon them (Kabeer, 1999, 2001b, 2011; Kishor & Gupta, 2004).

By the collective mainstreaming of women from mid-1990s, the strict sense of *purdha* has been replaced such that the practice of *purdha* does not mean always covering the body of women with *borkha*; it rather means appearing in front of male strangers decent and modestly covering their heads, going out wearing dresses and following the dress code. The recognition of women in the nation's constitution as equal citizens masks the overriding influence of Islamic religion on women's lives, which continues although outward form changes.

The practice of *purdha* differs as per the socio-economic background of women, education of the family, practice of religion in the family and the type of job of women and their husbands. In the coastal area, in the pre-*Aila* period, *purdha* practices limited women from being engaged with economic activities outside *purdha.* The respondents reported that this was an acute problem, particularly during disastrous times for those who strictly maintained it. However, this study, which has been conducted 7 years, after *Aila* (2009) does not identify strict consistency in maintaining the *purdha* system as reflected in earlier studies performed in the *Aila*-affected area (Baden et al., 1994; Cannon, 2002; Cunha, 1997; Ikeda, 1995; Kabeer, 1990; Khondker, 1996; Nasreen, 1995). This study reports that women noted changes during their lifetimes from being veiled and silenced to being free to move beyond the home.

This study reports that the introduction of different programmes by GO and NGOs has been influential. People have a more flexible attitude in the community and men and women are not strictly separated in the public and social sphere for everything. Women's ability outside the home allows women work in the fields and to be engaged in paid work and going to markets on their own; this is matched by their role in decision-making. For both formal and informal meetings for different projects and programmes operated by GO and NGO, women in the study area discuss social development issues, including family planning, small size of the family, its benefits, health issues, children's education, their welfare and future and women's rights.

Social safety programmes such as relief programmes have been reformed; they are no longer designed around religious observance. They are replaced by conditional cash transfer programmes, cash for work and a food for education programme (Ahmed et. al., 2008). Both VGD and VGF programmes target women in poverty and now include income-generating activities and training related to health, nutrition, family planning and education (Nguyen & Wodon, 2015). GO safety net programmes and other programmes provided by NGOs successfully reached vulnerable people in the rural area and assisted in changing patriarchal attitudes towards women. Consequently, the respondents report religious values did not create obstruction as they did in the past. When asked why do religion/religious values not create obstruction, discussants report that religious leaders are considered by GO/NGO as an important stakeholder in different safety net, training and awareness programmes. They are sensitised and aware via progressive advice about the benefits of changes in patriarchal attitudes for strictly maintaining *purdha*. In terms of sensitising and awareness of religious leaders about observance of *purdha*, mobility of women and rights of women, younger religious leaders are more tolerant than older/aged and they possibly accept modern views such as the reproductive behaviour of women and the *purdha* system, the discussants of the UDMC members opine.

Changes in practising *purdha* do not mean women totally abandon it; rather, it indicates women are no longer behind walls or fences and they are not confined to the house. Note that 73% of respondents of this study said that they do not maintain *purdha* in a strict sense. The *purdha* practice is dying in its typical sense because of the new economic contribution women are making to the household; it is a contribution that allows the family to have more nutritious food to eat in the coastal study area, which is consistent with Alston's (2015) study. This ability to provide food has given women significant new power in their relationships.

When asked about changes in socially established practices that occurred after *Aila* women discussants describe how they were unable to collect water or fuel wood from local sources and *borkha* made this extremely difficult, they had to change their practices. Women work outside and are involved with earning activities; they are now seen by strangers. Moreover, they have to work with male family members in the field; it is impossible to continue with *purdha* in a strict sense, i.e., not allowing themselves to be seen by outsiders and covering the body with a *borkha*. Women in female-headed households did not follow *purdha*. Some respondents report that because they live on the *wabda*, which is open, it is difficult to maintain *purdha*.

Saree is an extremely common dress for women in both rural and urban areas in Bangladesh, which women wear as part of *purdha* because it is a symbol of honour, tradition and modesty. *Saree* as a clothing is particularly catastrophic when disaster hits and people are unprepared or swept away by flood water. Women do not care about modesty when clothing is holding them back to save lives women took off their *sarees*, thus ignoring the dominance and prioritisation of women's "honour" (Alston, 2015; Nasreen, 1995). Ignoring dominance and honour reflect changes in attitude in terms of *purdha* and flexibility about maintaining *purdha*. This study reports that women travel in the same boat with males and works in the fields without veiling their face, which reflects changes in practices of *purdha* in the coastal area.

6.3 Involvement of Women with Earning and Outdoor Activities

The respondents report that coping with crisis is more important compared to maintaining *purdha*. Alston (2015) reports poverty and straitened financial circumstances are causing a revolution in women's working lives. Women are going away from their subordinate position to husbands and husbands' families and becoming busy by adding paid work and livestock-rearing works in the fields. Currently, many in southern coastal Bangladesh work outside the home, thus breaking down old customs and practices that confined women to the household compound. Therefore, *purdha* and male control did not stop them from doing their routine work for various reasons, including economic restructuring, economic development and financial crisis.

In terms of ignoring *purdha*, a woman respondent reported that breaking *purdha* practices and allowing women to earn money created certain problems, particularly for men. However, in the study area, all other discussants report that women were working outside because of circumstances and doing the work with the consent of male guardians of their family, including their husbands in coastal areas. The respondents report that changes in attitudes and local practices there have created a "mutual and cooperative" atmosphere and changed attitudes between men and women regarding the involvement of women with earning activities in the coastal area after cyclone *Aila*. They report that the involvement of women with income-generating activities indicates the overall status of women which has been improved in terms of their role in the decision-making process, gender violence and empowerment.

Sexual violence, abuse, health, injustice and inequality are evident in Bangladesh (Agarwal, 2001; Alston, 2015; Pakulski, 2004). For gender violence and gender-based discrimination, wealthier women of coastal areas are less likely to be victims of people outside their family because they do not go outside the home to earn money such as women of poor families. However, because of the flexibility of religious norms such as *purdha* for poor women and letting poor women work outside, the extent of discrimination against women and gender-based violence has been reduced in the coastal area. The amount of gender-based violence and discrimination against vulnerable women has reduced because of their awareness of their rights. Furthermore, they have the opportunity to file a complaint to the UP chairman's office and to the nearest police station, which has been contributed by GO and NGOs. To certain extent, NGOs help vulnerable women and NGOs organise women to complain against discrimination and gender-based violence.

Younger women now have an approach to avoid subordination because they have the opportunity to move outside the house to earn money, e.g., the opening of garment factories in Dhaka, Chittagong, Khulna and Rajshahi, has given them an acceptable alternative—one that provides remittance income for their families and allows them to escape the threat of early marriage and lifetime servitude (Alston, 2015). Importantly, because of additional flexibility of socio-cultural inequality such as *purdha* (which is grounded in structural constraints) in the urban rather than rural areas, women get access to some form of autonomy over their own lives, including the uptake of contraception, freedom of mobility/going out, decisions about children's schooling, the ability to gain control over various aspects of life and in defining aims and acting upon them (Kabeer, 1999, 2001b, 2011; Kishor & Gupta, 2004).

Though there are certain forms of discrimination against women, e.g., their wages are less, the difference has reduced as the involvement of women in earning activities outside the home has increased in the coastal area. To summarise, this study reports that because of changing attitudes and local practices towards women, the contribution of women in the family has increased and *purdha* has become more flexible and less onerous. The positive changes and involvement of women with income-generating activities gives both financial support and positively impacts other factors (Fakir, 2008).

6.4 Women Raise Voices for Dignity and Rights

Because women now have freedom of choice and women are now empowered to decide about themselves at both personal and public level in households, at a societal level, they have started talking/raising their voices for their dignity and for their rights in different platforms. Respondents report that the process has already started via social and economic change such as electoral participation, participation in civil services, participation in cooperative societies, participation with different public platforms of decision-making processes and participation in different movements. For example, women's representation has gradually increased in the Bangladeshi civil service. In 1994, the representation of women in the civil service was 9%, whereas it was 10% in 2002 (Jahan, 2007). However, in the last decade, the representation of women in the civil service has radically increased. In 2016, in the Bangladesh Civil Service Examination (BCS), 26.22% were women (Haque, 2018). Going out to work and adopting new practices are important sources of power of women; the opportunity of going out changes their attitudes (Ahmed et. al., 1997; Haque & Itohara, 2008; Rahman & Naoroze, 2007). Moreover, it makes them aware and conscious of their rights.

The consistent policy and programme of interventions by GOs and NGOs contributed to women empowerment and changing the gender discrimination in economic participation, in small business, in political participation and in reducing sexual harassment. For example, for reducing sexual harassment, sexual abuse and early marriage, both GO and NGOs established a "hotline service and one-stop service". The community women, the local school teachers, headmaster, UP members and community people have access to a one-stop service to complain anytime. Community women can directly complain about sexual harassment and abuse against anyone in the community. Sometimes, they seek help from a school teacher, headmaster or senior male member of the community to complain, which reduces sexual harassment and sexual abuse. The notable results show that the male members of the community, in the name of GO employees, NGO activists, school teachers and school headmaster, primarily play a role in empowering women in coastal areas.

Women's access to information in the name of "one-stop service" in coastal areas made women psycho-socially "empowered to". They have capacities including the capability to make decisions and authority to identify solutions to problems (Panday, 2016), which made them confident to move freely outside. Also access to information "empowered with" and "empowered within" (Panday, 2016) the women for collective action in the form of building awareness, consciousness, confidence building and self-esteem in the community about their dignity and rights. These changes have led to changes to the state of women participating in local politics and local elections, which is important for institutionalising women in democracy. It is reported in the study area that other than three reserve membership for women in the union *parishad*, women are seeking election for chairmanship of the union *parishad*, which reflects changes in political attitudes. Women are becoming more involved in public activities at a community level and find themselves in a more advantageous position within their communities but additional improvements are required as inequality remains. Carrying out the responsibilities of women, women representatives proved themselves to be honest and capable—even more so compared to men (Panday, 2016).

6.5 Involvement of Women in Disaster Management Process

Respondents report that women are involved with the disaster management committees and activities but additional involvement of women will give more positive results for reducing vulnerability. Alston (2015) proposes major advances can be achieved when women's rights, responsibilities and opportunities are recognised; when they are given equal influence on national priorities; when they are an equal part of decision-making on resource allocation and policy; and when they are integrated meaningfully into the work force with equal and living wages. One of the respondents report that it will take radical thinking to integrate women and make women leaders in disaster management rather than men because of socio-cultural practices. They argue that the existing long-practised norms cannot be ignored overnight in the coastal area. However, respondents argue that "involvement of women" or "leadership of the women" in disaster management has considerably increased in the coastal area. For example, members of the *Nari Bikash Kendra* bridge communication and relations among community women and GO and NGO to ensure improved service for vulnerable women.

Women's participation and involvement with disaster management processes reduces vulnerability and risk. The respondents report that by facing disaster repeatedly throughout life, women gain knowledge and skills that are extremely relevant to the culture of that particular society. The study results show that the high majority of respondents believe that the involvement of women with disaster management processes and development of leadership in coastal areas reduces the vulnerability of people in coastal areas. If more women are included in the UDMC, women will become more cautious about future shocks and they will be able to save additional assets and lives.

During a discussion, a woman says that there are males who consider women's involvement with development projects, disaster management processes and decision-making processes in the community is bad; moreover, certain male respondents say that the inclusion of women in the committee does not make any difference. Certain males agree that yes there are some males such as this but say it is not a general scenario of the community for the last few years and add that the overall situation in the coastal area has changed after *Aila*.

The inclusion of women because of rules and maintaining the order of GO (which is defined as *niom rokkha*) is not important; rather, the effective inclusion of women to ensure equality in decision-making processes related to community interests is necessary. Certain women report that women's representation must enable women to raise their voices and share their experiences, which will then help make future policies and plans because they have an improved understanding of women's requirements (Panday, 2016). This study reports that along with different training and awareness programmes, NGOs keep women informed about important national and international issues and not only women-related news that is published in daily newspapers. This helps local women establish a network and keep women aware of the world outside their community. In Bangladesh, without linking women at local rural levels with national-level policy-making, activism of women will remain locked in disempowered positions (Alston, 2015). However, respondents argue that socio-cultural practices of the community cannot be ignored overnight and radical changes must be carefully introduced.

This study reflects that community-centred and collective warning systems considerably reduce the loss of lives and assets. Respondents report that, in the coastal area, a mass of people and community-oriented warning system have developed in which the primary focus is to involve the community people, particularly women. Consequently, despite the same magnitude cyclone *Sidr*, a category four storm, killed 3,369 in 2007; only 190 people were killed by *Aila*; however, in 1971, the cyclone Gorky, which was of the same category, killed 139,000, which is considered to be the result of effective forecasting (O'Donnell & Wodon, 2015; Paul, 2009).

The development of effective warning and communication for disseminating information required strong technical foundations and good knowledge of the risks. In the coastal area, Bangladesh has developed strong technical foundations such as television, mobile phone/networks, community radio and local disaster management committee (UDMC, CPP, PIC). The technical foundation in the coastal area results in gaining better knowledge and mobilises local community people, particularly marginal people, about risks and vulnerability related to disaster. All respondents report that the networks for disseminating warning for hazards are effective, and networks have been successfully developed because of the development of strong telecommunication systems and technological advancement such as mobile networks. The advancement of mobile networks all over the country, even in remote areas, played an important role for disseminating warnings, as well as reducing damage, vulnerability and risk to people in coastal areas.

The development of networks and communication and "working together attitudes" led to a paradigm shift in disaster management from a conventional response to a comprehensive risk reduction culture and strengthens the capacity of coastal people in disaster management. The development of technology, additional involvement of community people and consideration of cultural sensitivity issues while disseminating information play a role in developing community-centred and collective warning systems in the study area. This study shows that the availability of the warning information to women created opportunities for women to disseminate to other women, which resulted in easier evacuation and reduced damages and deaths.

6.6 Summary and Dynamics of Discharging Vulnerability (DVM) of Women

In this study, people who participated are certainly aware that there are significant changes in the traditional roles and responsibilities of women that include changes in socio-cultural, religious and economic roles. Women are more aware than ever of their rights and they are more conscious about their role in the economy, decision-making processes and policy level, particularly after cyclone *Aila* in May 2009. Transformation in the coastal area has started, although there is still a long way to go. There is gender inequality but changes in position, roles and responsibilities of women have certainly progressed in rural Bangladesh.

Transformation in the coastal area includes changes in the everyday life of women, including flexibility in *purdha* and religious practices, mobility of women, working outside, raising voices for honour and rights in different platforms, involvement with disaster management processes and access to resources such as income. The social transformation and change include ability to take challenges by the women and empowerment of women in terms of their political participation, economic participation, participation in decision-making, literacy rate, flexibility in socio-cultural practices (summed up in Fig. 6.1). Moreover, these changes developed the agency of women more strongly because of policy implemented by GO and NGOs. However, additional results will be reflected if policy and programmes are implemented in an effective manner, e.g., for ensuring the representation of women in a meaningful way at every stage (Fig. 6.1). Although complete social mechanisms changed in favour of empowerment of women, women of the study area remain vulnerable for structural, social, institutional, economic and for cultural reasons. For example, the lack of transport communication, secured shelter centre and lack of a suitable source of drinking water near the community, make women vulnerable. From the PAR model, Hegemonic Masculinity and Feminist Political perspective, the social, political and economic, educational, infrastructural and policy factors impact the disproportionate vulnerability of women. However, from the perspective of Access to Resources and

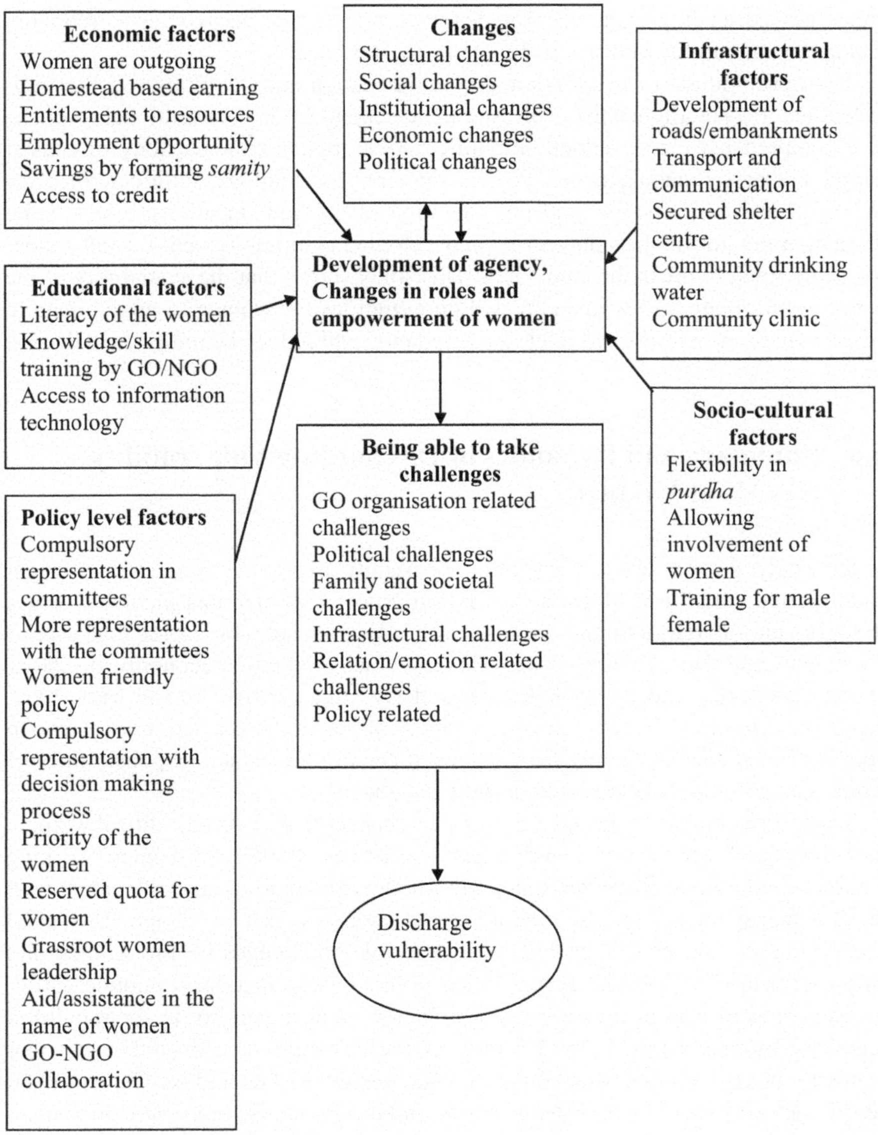

Fig. 6.1 Dynamics of discharging vulnerability model (DVM)

Entitlements and Social Transformation, letting women go out to work, ensuring their entitlements and access to resources, education, participation in politics and implication of the initiatives correctly makes women confident. Furthermore, it enhances their self-reliance to take challenges and to develop agency; these ultimately reduce vulnerability.

Chapter 7
Implication of the Findings, Conclusion and Recommendations

7.1 Introduction

The empirical results of the study have a theoretical implication for addressing disaster, vulnerability of women, transformation and changes in coastal areas. The results suggest that intensity, livelihood, likelihood, speed of onset, familiarity with the disastrous environment and consequences of Cyclones such as *Sidr*, *Aila* were considered as significant hazard characteristics to the participants of this study. This study reports that the affected people had moderate ability to reduce the risks of a hazardous event. This study develops fewer models and aspects to go beyond and increase the analysis of disaster, women and vulnerability in a traditional coastal area that has not been addressed in other studies. The models are, namely, progression and pathways of vulnerability model (PPVM: Fig. 3.9), coping and adaptation model (CAM: Fig. 4.3), dimensions of empowerment and change in the role model (DECR: Table 5.1) and discharging vulnerability model (DVM: Fig. 6.1) of women based on the idea of PAR model of vulnerability, access and entitlement to resources model, feminist political ecology and social transformation theory, which examines root causes, dynamic of pressures, unsafe conditions of vulnerability, mechanisms of discharge, social transformation, social changes and various elements of disaster risk. The following discussions link the results of this study with theoretical and practical implications to facilitate the understanding of disaster, women, vulnerability and changes in the Bangladeshi coastal area.

7.2 Theoretical Implications and Contribution

7.2.1 PAR Approach and Vulnerability in the Study Area

Along with PAR as a theoretical consideration, this study demonstrated that vulnerability of women to increasing disasters is largely affected by their economic,

M. M. Khan, *Disaster and Gender in Coastal Bangladesh*,
https://doi.org/10.1007/978-981-19-3284-7_7

educational and socio-cultural background, in addition to existing infrastructural factors, policy and socio-cultural factors. Saha (2015) assesses the vulnerability of the southern coastal people in Bangladesh using PAR and explores the influences of various socio-economic, environmental, institutional and geographical factors on escalating risk and vulnerability. Across coastal villages in Bangladesh, the risks of a hazard event are similar which has been figured out using DCM by Saha (2015). Disaster risk theory explains that structured and unequal vulnerability because of unequal access to resources, lack of entitlements and socio-cultural factors create additional disaster and risk for certain groups of people. As per the PAR theory, the increasing pressures resulting in vulnerability and pressure arise from the discriminatory economic and social structure that affects resources. The PAR model explains that socio-economic and political causes are the root causes rather than geophysical and biological factors for vulnerability of women in coastal areas. To address vulnerability of women in the coastal area during a disaster, the research question addresses local practices that are causing disproportionate vulnerability for women. In response to this research question, socio-cultural barriers such as religious barriers, patriarchy, hegemonic masculinity, marginalisation of women, the absence of male family members and the dependency of women on men are identified by respondents in this study as pressure for certain groups of people such as women, female-headed households and poor people; moreover, these pressures create disproportionate vulnerability.

However, socio-cultural and political pressure such as political ideologies, corruption of local leaders, less representation of women in different committees, and keeping women away from decision-making process can be a cause of vulnerability. Practical policies can be formed from PAR studies, e.g., poor people are more vulnerable because their housing is in unsafe areas. People who take shelter under open skies and safe shelter centres socio-economically are insolvent. Therefore, some of those are sheltering with rich people in the community. If they are not assisted to build brick or other stronger houses above the flood prone areas, mobility to other areas of these people and downward economic mobility of these insolvent people will continue during a disaster.

Hazards such as *Aila* took away houses and wealth of rich people. Their limitations differed from those of poorer people because rich people felt embarrassed to go for aid and assistance but they did recover much more quickly. In the policy, to reduce vulnerability rather than taking strategy in general, strategies to reduce vulnerability should be specifically addressed from different aspects such as economic strategy, cultural strategy, political strategy and geographical strategy. For example, to reduce vulnerability because of not having a house, "a one room house" built by GO has been provided to those who lost their house because of *Aila*. However, sleeping in the same room with adult children pushed their parents into culturally vulnerable conditions that require to be addressed in the policy. Saha (2015) concludes three years after *Aila* the villages studied were suffering from significant vulnerability that they "had moderate ability to easily reduce" (Saha, 2015: 727). He called for pragmatic strategies and for the state to address "fair allocation of wealth, development policies, addressing corruption, marginalisation and pauperisation that are root causes of vulnerability" (Saha 2015: 749).

7.2.2 Access and Entitlement to Resources, Disaster and Women in the Study Area

Access and entitlements to resources involve the ability of an individual, family, group or class at normal or at disastrous times to adapt to have ability to deal with threatening situations, to have control and to have command of the totality of rights and opportunities. Wisner et al. (2004) report unsafe conditions in society arise because of economic and political process that allocates assets, income and other resources in society. These unsafe conditions are variable, disproportionate and unequally impacted on a certain section of people in society as well as the unequal distribution of resources in rural communities. Access to resources model explains that vulnerability varies in society because of unequal distribution and access to resources, which depends on gender, class, caste, income and employment opportunities (Wisner et al., 2004). Sen (1981a) reports that entitlement to resources of marginalised, under-privileged individuals and groups is denied or lessened because of their inability to command the legal means available in the society. Sen (1981a) argued disproportionate vulnerability results from a failure to be entitled to any bundle of rights, including the availability of food in the society. Considering this theoretical explanation, a research question of how does unequal access and entitlements to resources impact women in vulnerable situations? This study reports that in the study area women were lacking access and entitlements to resources that pushed women in the coastal area to disproportionate vulnerability, particularly before *Aila*. Respondents and discussants report that women did not own rights to resources and properties such as male family members in the coastal area. Because of the lack of entitlement and access to resources resulting from socio-economic and political process, women were not even playing an important role in decision-making by themselves; rather, most decisions were taken by male members. The unequal distribution of resources in the community keeps women powerless and marginalised materialistically and psychologically and that often increases vulnerability (Blaikie et al., 1994).

After *Aila*, because of the realisation about the importance of women's participation by the GO at a policy level, access to resources and information systems have been ensured, and the consequence is the enhancement of women's agency and coping and adaptive capacity. The respondents report that access to resources, access to micro credit, and employment opportunities socio-psychologically boost women and promote self-esteem, self-efficacy and psychological well-being. Furthermore, this study highlights the social acceptability of moving women outside the house is dynamic rather than a static patriarchal mentality issue for the entitlements of women to resources. Vulnerability has reduced irrespective of whether the woman is in a wealthier or poorer family in the study area, particularly after *Aila*. There are no existing policies of GO and NGO that directly address access and entitlement to resources; rather, they address mainstreaming women, providing training and raising awareness and ensuring participation of women in earning activities. This is the only academic study in Bangladesh that researcher is aware of that used this approach where it is revealed in the study that both GO and NGO policy-makers should include the issue of access and entitlements to resources for mainstreaming women rather than mainstreaming through providing training and raising awareness.

7.2.3 Feminist Political Ecology and Changes in Coastal Areas

Feminist political ecology theory considers the relation between gender and environment because of the social construction of gender, class, geographic location, employment opportunity or lack of opportunity and economic difficulties. Feminist political ecology examines gender as a factor to development because the relationship between gender and environment determines access to and control over natural resources. Feminist political ecology approaches that men and women are differently affected not only in unequal political, economic and socio-cultural structures but also in the ways men and women differently relate to nature and environment. The differences in relation to nature create differences in responsibilities and experiences of women. In relation to this argument of feminist political ecology, this study reveals (which has not been used by any other researcher to study the vulnerability of women in coastal area in Bangladesh) that access to natural resources and environments varies based on gender in the coastal area and shapes the structure of the economy. Because of the lack of access and control over resources, women are overlooked and excluded from the management and decision-making processes at both family and community levels in the study area. This impacts the way they differently experience the environment, e.g., women from poor families live on abandoned land in close proximity to the river. Because of the location of the houses in close proximity to the river, they have no permanent shelter and always have to move; furthermore, frequent mobility escalates poverty and inequality to get access to natural resources, as revealed in this study.

Simultaneously, they face unequal discriminatory behaviour by local people. Because of gender roles and responsibilities, women do household-related activities, including cooking, water collection and fuel wood collection (non-monetary/non-productive work), and they are more affected and contaminated by polluted water. Furthermore, women who work in the *gher* and in agriculture, which are run and primarily controlled by men, are being exploited by getting lower wages compared to men. Gender-based division of labour, discriminatory material practices, distribution of property and power determine the interaction with nature and natural resources (Nightingale, 2006; Nonuguchi, 2012).

Women are overlooked and subordinated in the coastal area in Bangladesh by patriarchy and market economy (capitalism) as nature is overlooked and contaminated (Mies & Shiva, 1993; Rocheleau et al. 1996). The access and control over natural resources are determined in the coastal area by gender, responsibilities and power. Women respondents describe the lack of access and control over natural resources in that they are not getting access to Sundarbans to collect honey and wood only because they are women. Nevertheless, if women could have access to Sundarbans they could collect honey and wood, which would empower them financially and allow them access to market. It is not only the family and community denying women's access to Sundarbans but also the administration who issue permission to get into Sundarbans.

The inequitable division of labour, responsibilities and power structure has been reduced but not completely eliminated. Discussants argue these unequal impacts will not be removed as long as women lack access and control over resources. Giving freedom, valuing privacy of women and the principle of equality between men and women socio-culturally reformed the coastal area. Women are protecting and conserving the environment as both participants and partners in environmental protection and conservation programmes via plantation around residences and performing household gardening. The respondents acknowledge that women earn money and contribute to the family by gardening. Furthermore, they are planting wood and fuel wood trees, which protect their houses from strong wind and storms and produce oxygen and provide fresh air for a healthy life. In terms of tall wood trees, bees build their nests and some women are farming bees for honey in the trees, which makes them financially empowered. Discussants note that while males usually chop down the forest, women care for the forest. It is revealed that when men get access to Sundarbans, they randomly chop trees for fuel wood and destroy the natural setting of the forest to collect honey and go fishing. However, if women could get access to resources, they would collect wood and honey using their motherly attitudes and experiences. A male respondent reported that women are after all caring. Gender relations, gender roles and power structures shape environmental decisions, laws and access to and control over resources, and the gendered knowledge and gendered environmental rights should be addressed by policy-makers.

7.2.4 Social Transformation Theory and Changes in Coastal Area

Social transformation theorists (Beck-Gernsheim et. al., 2003; Giddens, 2009) report that solidarity, equality, expansion of facilities, distribution and ensuring opportunities for all (infrastructure, communication systems, technology, access to technology, transformation of information from one individual or group to another, and formation of networks) minimise distances towards women. This helps with increasing awareness and consciousness against the oppression of the women. Social transformation promotes activism and enforcement of policies at a grassroots level that has led to desirable outcomes for rural women in Bangladesh (Sultana, 2011) and challenges existing structural inequality (physical, economic and cultural) and brings sustainable change in society (Watkins et.al., 2012; Noor, 2016; Panday, 2016). In the study area that the ability of women for the social transformation of their environment makes women able to face existing challenges better and utilise their capacity and potentiality better. They receive additional respect from family members, increase mobility and additional freedom increases their self-esteem and bargaining power at both family and community level. Changes occur because of pockets of activity by both GO and NGO whose vision of an alternative form of social organisation is more altruistic and cooperative (Watkins et. al., 2012).

Women's voice is being heard and self-esteem and bargaining power of women are evident in the study area because of the emphasis on mainstreaming women in different policies (e.g., SOD, ccGAP, BCCSAP, NAPA) that is formulated by GO in Bangladesh. In addition to these policies, Bangladesh GO (2011b) has enacted a particular policy entitled "National Women Development Policy – 2011", to mainstream women with the development process. Furthermore, in Bangladesh, the movement for women's emancipation since the 1980s by different NGO organisations, civil society organisations and women's organisations framed the basis of women's political voice and activism, as well as formed the foundation for women's human rights.

Moreover, technological advancement and access to technology (satellite TV, mobile and internet) have instantaneously and simultaneously facilitated the connection of women in the study area with people outside the community, which is important for changes in the coastal area. They have connected women with communication, dissemination and network, as well as facilitated the development of a community-centred and collective warning system where women are included. For example, as shown in this study, the result of the inclusion of women in the warning system facilitated improved management and resilience during a disaster.

The mainstreaming of women is important for any transformation in the society. Bangladesh has become a model for women's empowerment; it is listed as one of the ten "fast-track" developing countries in the world. Moreover, Bangladesh is ranked as the first in gender equality among South-Asian nations because of significant progress in maternal and child health, success for 100% enrolment of girl's education, and 60% of primary school teachers are women (World Bank, 2009; WHO, 2012; Khaled, 2018; UNDP, 2018). In Bangladesh's garment sector, 85% of workers being women now accounts for 80% of Bangladesh's total foreign earnings (CRI, 2018) and bringing women under the Maternal Health Voucher Scheme has strengthened health services, reducing fertility and mortality (World Bank, 2009). Moreover, these achievements were not by any means accidental nor did it happen overnight; rather, these achievements are a consequence of social transformation by mainstreaming women in Bangladesh, including the study area. For example, the mainstreaming of women in the coastal area functions as a safeguard in terms of access and control over resources and participation in community decision-making, which played an important role in building resilience and capacity across the household and community levels (WHO, 2012; UNDP, 2018).

Furthermore, the social transformation facilitates the development of the agency of women. The developments of agency of the women facilitated to have control and access over resources as well as changes in the everyday life of women. As the process of development of agency, women in the study area run small businesses, work in the fields, fishing the rivers, construct dams, campaign for election accompanied by males, and go out without doing *purdha* strictly. These transformations taught women to take challenges and help them speak up, to listen and to stand up on their own for their dignity and rights.

This study suggests that disaster as an event/a factor contributes to bringing changes and social transformation in a society such as Bangladesh where there are long existing social practices. The feminization of role and responsibility of the women and gender relations in the coastal area has changed in the last few years, particularly after *Aila*, in addition to the advancement of technology, economy, social, cultural and religious practices. Although small in number, male members collect water and fuel wood for women, women work with men in fields, *gher* and fishing in river. This indicates the *purdha* system does not restrict women. This social transformation in the eyes of coastal people happens more because of the self-realisation of male members in the community, recognition of women's contribution, necessity and scarcity after *Aila* compared to alteration in the social structure.

7.2.5 Patriarchy and Masculinity and Its Changes in the Study Area

Gender and changing nature of gender relations are determined via a socio-cultural process in which an individual cannot keep himself or herself separated but these are constantly in the process of being remade by women being accepted in politics, economics and decision-making (Butler 2004; Connell, 2005; Kimmel, 2008; Kabeer, 2011; Panday, 2015). Connell (2005) reported that gender and gender relations are developed by social interactions. In this study, the concept of masculinity and patriarchy has generated research questions to explore gender discrimination, gender exploitation and gender dominance of male in a coastal area. Inequality between men and women, hierarchy, power dominance and hegemonic masculinity was not fixed; rather it is amenable to be changed, eroded and culturally and structurally replaced, which is revealed among people in the coastal area.

Respondents report that uprooting masculinity and patriarchal domination requires time. However, after different interventions by GO and NGOs, the dominance of men in coastal Bangladesh has been reduced, e.g., respondents argue that the migration rate of women from the study area is smaller compared to male members.

Nevertheless, in terms of the women who migrate, a small number of them are fundamentally influenced by the socio-cultural context in which people are embedded. The rate of migration and participation of women in the labour force has increased because of free education for girls, priority and quota of women in employment opportunities, priority for women in different GO and NGO projects and programmes, micro-credit loans for women, loans for small entrepreneurs in rural areas and the growth of the garment industry. This leads to the socio-economic empowerment of women.

Furthermore, this study uncovers that flexibility in masculine and patriarchal systems empowered women to social, political and educational opportunities, which impacted corruption and sexual harassment. Traditionally, in Bangladesh, the practice of patriarchy was dominant in every step, although Islamic law, *shariah* and the constitution have recently given rights to women. The inheritance of patrimonial properties to women was not being followed when first enacted by GO; rather, they were violated by the patriarchal and masculine social systems (Kabeer, 1990). Nevertheless, for the improvement of society, women now have control and entitlement to properties (Kabeer, 2011; Sayeed, 2010).

This study reports that the involvement of women with the distribution of aid reduces corruption and bias for distributing aid. This is because of mandatory inclusion of women within different committees at local and community level, in addition to the increased tolerance of male members to women's participation for community-based decisions (Panday, 2016). Such mainstreaming, mutual trust, understanding and respect between men and women on the grounds of gender equality have allowed women to participate in public affairs, which are a reflection of flexibility of masculine attitudes. At present, religious leaders accept that, for changes in the community, women's participation is required because women have an improved understanding of women's requirements. Furthermore, because of the involvement of women with different committees and mainstreaming of women by GO and NGOs, women are aware of their rights and they do know where to complain and ask for justice. The participation of women in decision-making platforms is evident, thus making their "voice heard" by male members of family and community. Participation and awareness built their capacity and developed agency about legal rights and legal support by GO-NGO against violence and sexual harassment, which has been decreased.

This is demonstrated by respondents and discussants saying that no women are wearing *burkha* to participate in FGD. Is this not a change? Women are working and involved outside with income-generating activities. Is this not flexibility? Women are attending meetings of different committees related to disaster management and training programmes provided by GO and NGOs. Is this not the breaking/changes of male dominance? All concluding remarks have the potential future implications for policy and research considerations. The next part of this chapter focused on implication issues with limitations.

7.3 Policy Implications

The results of this study suggest a number of approaches by which policy and programming could be improved to cultivate better gender equitable attitudes in the coastal area in Bangladesh. The prevailing negative attitudes of the coastal area that reduce women's self-esteem require additional change after the positive attitudinal changes that have already been found in the coastal area towards women because of different motivational programmes by GO, NGOs and the community itself. The projects in future should prioritise a bottom-up approach with the voices of the local people ensuring gender neutrality before starting any projects and programmes for people in the coastal area. Along with mainstreaming marginalised, under-privileged and community people, there has to be coordination among policy-makers, GO, donors, NGOs and the local population. For example, the discussants reported that creating employment opportunities in the coastal area will provide a better opportunity to manage scarcity compared to providing loan and relief. Moreover, political empowerment and ensuring access and entitlements to resources by women will uplift women's self-esteem and improve their awareness of their rights rather than provide aid and assistance in the name of women.

Discussants argue that considering the political empowerment, entitlements, and mainstreaming women in decision-making platforms issues by applying a bottom-up approach for hearing the voices/experiences of local coastal people by policy-makers and project implementers must increase the effectiveness of implementing aid and assistance programmes after disasters. In coastal areas, there are certain long-standing socio-cultural practices that cannot be ignored and changed overnight in a radical manner. Certain male members do not allow women to work outside and certain families are financially deprived by not accessing GO and NGO projects, thus reflecting the requirement for further change and more sensitivity to local social practices by policy-makers.

For empowering and developing the agency of women, GO intervenes via various social safety net programmes such as VGF, VGD, widow's allowance and aged allowance; however, they are not as successful as NGOs. GO is unsuccessful for ensuring safety programmes that empower women because of the lack of good governance, bias and bureaucratic complexity of administration, lack of access to the individual at grassroots level, ignorance of the local people by GO in social safety projects, corruption, misconduct and bias of GO officials. Furthermore, the programmes of the GO are primarily functional at a macro level and policies and projects are initiated from the top, and the stakeholders from a grassroot level are not involved. The voices of women in the community are not reflected in the GO's decision-making process after a major disaster; rather, the voices of the powerful and influential male members of the community are heard and considered with care in the decision-making process even if the decision will negatively impact women. GO planners and practitioners are not aware of the real scenario of vulnerability, adaptive capacity and unequal settings of the coastal community because they operate their programmes only during a crisis and they only visit the locality on an occasional basis.

However, NGOs have picked up the functional weakness of the GO and ensured the implementation of different programmes for women such that they can control their lives and they can have capacity to make decisions. NGOs directly work at a micro level with basic demands of the grassroots, vulnerable and marginalised people. Therefore, because of the nature of activities, women are given priority by NGOs in different programmes. The access of NGOs to the grassroots level without following administrative complexity, flexible accessibility and soft power practices made their visibility higher than GO at grassroots level in the coastal area. Because NGOs deal with grassroots people, primarily women, NGOs are more transparent, accountable and responsible for their programmes after the continuation of NGO programmes depends on the success of their existing programmes. Moreover, NGOs work in the coastal area throughout the year and they deal with people at grassroots. NGO activists have better experience because they deal with the local people/local area than GO officials and NGOs are more transparent and accountable for their role and responsibility. However, both GO and NGOs have the opportunity for training but GO officials are more indifferent and less accountable. Although the participation and representation of community people from all stages, including women, in different disaster management processes and committees for all in Bangladesh (GO, NGOs and local authority) has been ensured by SOD, ccGAP: Bangladesh, BCCSAP and NAPA; the number representing marginalised people such as women is still poor compared to the representation of male members.

Discussants argue in this study that despite the official representation of women, their voices are ignored by male members and most of the time male members do not even ask for the opinion of women members. Therefore, it is important for policy-makers to increase and ensure good number of women representatives. The representation of women in the disaster management process and policies should be taken based on the voices of all stakeholders, beneficiaries and vulnerable people. Finally, coordinated and concerted efforts among GO, NGOs and community people are certainly required for developing sustainable disaster management in coastal Bangladesh. For this purpose, a monitoring team, composed of GO, NGO and community people, follow up implementation and coordination of policies among GO and NGO with both men and women promoting young leadership, training for community male members about positive aspects of women's representation could have been performed.

NGOs created opportunities for women in political empowerment by promoting their direct contention in local elections. The political empowerment of women has resulted in automatic changes in the social structure and social value, including at family and community level. The socio-cultural grounding for women's empowerment is nurtured in a better manner by NGOs prioritising and mainstreaming marginalised people at the grassroot level. The respondents and discussants suggest that both GO and NGO should particularly focus on the gender-based requirements to mainstream women in the disaster management process. For example, certain programmes help men realise that household work is for both and it increases the role of women in decision-making power at family and community level. One of the major problems the respondents and discussants identify is that because of the workload and household responsibility, women cannot attend meetings of different committees. However, if someone else, such as the husband, shares the household chores, participation of women in different meetings and training provided by different organisations will increase and the bargaining and adaptive capacity of women will increase.

In terms of changes in attitudes, policy-makers and practitioners should organise a rehearsal programme on roles and responsibilities in disastrous times at grassroots level. Although SOD ensures rehearsal programmes on the responsibilities of the different national and local committees for disasters, respondents and discussants opine there has never been a rehearsal in the coastal area. People do not always follow policies and strategies in policy papers. When young, educated people may not trust the structure of disaster prevention and warning systems for a safe evacuation in the policy papers. However, if they do a rehearsal practically it will be effective to inform the coastal people about the devastation of the cyclone and the consequences of devastation will be reduced. The discussants argue that certain senior community people have already experienced cyclones two or three times in their lifetime or even more. Nevertheless, cyclones such as *Sidr* and *Aila* do not happen every year and community people do not regularly move into shelter centres for the events. Finally, this study reveals having one's own experience by facing cyclones about disaster prevention and warning for safe evacuation is a not professional experience. The study reports that practitioners who work with people at a grassroots level can easily convince them by showing and explaining the details of the actual functions.

One of the major challenges for policy-makers is to reach the poorest of the poor at a grassroots level; those people who basically have no land of their own to cultivate. They live in abandoned land and on dams or *beribadh*. Although their children have access to education and education is for all, most of their children have dropped out; they start working as child workers or as child labour. They are the least self-motivated because they believe themselves to be powerless and helpless in the community. The poorest section has no resources of their own to use in disastrous times for adaptation; furthermore, day labouring is their only source of income, which has been shattered in disastrous times. However, providing the necessary finance with logistic support to settle their residence in a permanent place will be helpful and will increase their income and self-esteem. For example, according to the discussants, along with relief and other allowances, arrangements for a small business in the local market or equipment for fishing will be a better support for the poorest people.

To reduce disproportionate vulnerability of such women and improve their adaptive capacity, they are already mainstreamed by providing a house in a permanent place, known as a model village. Furthermore, their obligatory representation has ensured aid and assistance, participation in food for work, food for cash, training programmes, social awareness programmes and political empowerment programmes by both GO and NGOs, although still they face challenges and obstructions. The formation of cooperatives, self-group or committees at a grassroots level only by women will play a more significant role in disaster management and reduce vulnerability in the opinion of the discussants. Both men and women discussants prefer that in addition to the combined committees of both men and women at local level, certain committees should be formed comprising only women, and their ideas should be adopted by mixed committees and that will break the traditional male-dominated top-down approach. This will help rural women to come together to act as a pressure group.

Gender programming focuses on gender discrimination and inequality on a regular basis in the coastal area but respondents and discussants consider that training programmes operated by NGOs in the coastal area are primarily for women. If the programme organisers ensure the presence of both men and women, training will help reduce family disputes. Discussants argue domestic violence, sexual violence and harassment are related to men's traditional authority over women. Addressing women's requirements and ignoring men increases family chaos, conflict, and family violence because men believe that they are losing traditionally imbued gender power in their relationships. Ensuring participation will significantly encourage men to become more gender equitable. Gender-oriented programmes certainly contribute to the accumulation of experiences and drive men towards gender equitable attitudes.

The study reveals that women in the coastal area have no access to and participation in certain social programmes, including village arbitrations, village meetings and voluntary activities. Though there are female-elected representatives in the local area, the elected women are not welcomed and invited to attend these meetings even those which are related to sexual violence, harassment and family disputes. However, discussants argue participation of women with different action-oriented discussion and village arbitration in the coastal area will improve the leadership skills of women. Women's leadership skills will certainly strengthen rural women's cognitive domain and bargaining power at both family and community level. Policy-makers should take initiatives to ensure the representation of women in these social programmes, including village arbitration only as there is mandatory representation at different disaster management committees.

Moreover, certain respondents argue that women have difficulty fitting in their double responsibilities both outside and inside the home. Ways to face problems, techniques to manage time and involvement with income-generating activities varied from woman to woman. Women who are aged, unmarried and with young children preferred to do income-generating work close to their home or inside their home and the middle-aged, married women and older women engage in work outside the home, including agricultural work. In terms of the distance of work, certain participants consider that many women work five miles away from home and walk all the way because there is no mass/public transport in the locality other than motorcycles and trawlers. However, the number of motorcycles is not sufficient and they are extremely expensive; the trawlers only run on rivers or canals. Women respondents appeal to both GO and NGOs to consider the distance before organising programmes and initiatives suitable for them.

Men should come forward for household activities and help women in household chores as it brings improved understanding of what women achieve. Women respondents demand that, for sustainable development and empowerment of women, separate training programmes should be undertaken in the study area. To some extent, male respondents report that empowerment programmes for women will be effective if male members of the community allow women to go out; most of the problems will be solved if this obstruction is resolved. Men realise the importance of the leadership and management capacity of women in the coastal area for disaster management. Both GO and NGOs should undertake programmes in which both men and women in the community participate to make them understand each other. An improved understanding between men and women will resolve problems better, empower women and ultimately help in successful adaptations and sustainable resilience during disasters.

7.4 Research Practice Implications

In coastal areas, disaster, women and vulnerability discourse is not an easy issue to examine in the context of socio-cultural characteristics such as gender relations, access to resources, religious practices, employment/unemployment, roles in decision-making, and political empowerment in the study area. However, this study confirms the strength of mixed methodology as a form of triangulation, particularly for dealing with gender-sensitive issues and makes it possible to complete such a gender-sensitive study. The mixed method has been used to contextualise the gender relation and behavioural pattern towards women, including sexual violence/harassment, discrimination, exploitation and ignorance of women in disastrous times in the coastal area. For quantitative data, the survey questionnaire for face-to-face interviews was divided in different sub-sections such as socio-demographic characteristics, disaster experience, gender role and vulnerability, entitlements and vulnerability, disaster and women's leadership, GO and bottom-up approach, disaster and role of NGOs, and adaptation strategies and resilience by women. Research assistants were asked to ensure respondents knew there was no pressure to answer questions; the respondents could withdraw midway through the interview if they wished but that did not happen primarily because of the good relationship between the respondents and research assistants. In face-to-face interviews, research assistants attempted not to be disruptive of the respondents' everyday lives and talked with respondents wherever they were working.

The survey study misses certain contextual issues related to sexual attitudes and perceptions of male members in disastrous times, which this study attempted to discover via qualitative means. The survey study was individual face-to-face interviews that provide experiences of particular individuals. However, sometimes, individuals ignore certain personal experiences, particularly issues related to sensitive and private/personal issues such as sexual harassment, menstruation and male dominance. These could be talked about in the context of normative behaviour in their community, which became possible by focus group discussion and in-depth individual interviews.

For qualitative data, focus group discussions and in-depth individual interviews with people from different sections of the community such as female-headed households, vulnerable men and women and different committee members made participants feel at ease as they were surrounded by members of the same community and same experiences. For example, FGD with vulnerable women and female-headed households and in-depth individual interviews made women confident to share sexual behaviour-related issues. Moreover, FGD with a mix of vulnerable men and women gave the opportunity to cross-check and justify the argument from both men and women.

Sometimes, even in the focus group discussions, discussants did not share their own experiences for either shame or fear of known and powerful people who might hear about them, e.g., they did not speak about their experiences of sexual violence and harassment. The in-depth individual interviews with women by female interviewers (research assistants) made women interviewees feel more comfortable and at ease for sharing sexual subject matter, abuse and discrimination. Although information of the women respondents differed from that of the male individual interviewee to some extent. Note that mixed method is a good medium for cross-checking and therefore for ensuring maximum reliability and valid information in terms of gender-sensitive issues. Furthermore, methodological triangulation works better for discovering the discourse of social realities in daily practices.

In a traditionally conservative society, building a rapport and getting access was extremely difficult; however, at the start of the field visit, particularly during the pilot study, the villagers were not as friendly as researcher found them at the end of field data collection. To some extent, the local UP chairman and members were not extremely friendly and viewed fieldwork with suspicion because some information was directly related to their activities. For example, while asking about the complaint of a beneficiary regarding corruption as well as bias in the distribution of aid and assistance, a UP member immediately declined to respond to the allegation and sometimes said they do certain favours for their own people. To discover such issues, in-depth survey method, FGD and in-depth individual interviews have been conducted in this study. Thus a rapport was built with community people because researcher spent six months there doing fieldwork. It was not easy to leave the research setting at the end of filed survey as researcher had built up a friendly relationship with some of the people of the area.

7.5 Implications for Future Research

The people from coastal areas live in a low-lying region and are extremely familiar with hydro-meteorological hazards such as tropical cyclones and storm surges. The study villages of Sutarkhali and Kali Nagor are extremely close to the Bay of Bengal and are extremely vulnerable to natural hazards, particularly tropical cyclones that occur almost every year to a small or large extent (Saha, 2015). To some extent, the study villages are not at all similar to villages in other parts of the country, and they even differ from other villages in the same area because of the geographical location. Some of them are detached from the mainland of the locality and consequently there is no direct walkway/communication.

The results of this study have been compared with other research in different areas. The other areas are different from this study area not only because of the different socio-economic background and the geographical location. The socio-economic difference of the study villages includes employment opportunities, alternative employment options, the bazar system, the communication system and transport.

In the coastal area of Bangladesh, transport-related communication is not the same as in the study area. The direct communication of the coastal area depends on whether the area is separated from the mainland or not. If it is separated, the way of transport communication is either walking or by boat and trawler; however, if the coastal area is not separated from the mainland there is better transport communication. The geographical location of this study area is quite different from other villages in that the study area is extremely vulnerable to natural hazards. Both GOs and NGOs could do large-scale research because they deal with large number of community people within their different programmes.

This study focuses on women and female-headed households on an extremely small scale. However, additional study can be undertaken by policy-makers, policy planners and academicians with more time to allow better insight for specific women's groups such as female-headed households, separated and divorced women, and widows to understand the dynamics of disaster and gendered vulnerability better from different aspects. The interviews and discussion related to the unequal distribution of risks, unequal access/entitlements to resources, women, and vulnerability provided an opportunity to the respondents, discussants and community people to realise differences among different groups of people in the community and even among members in the family. The research questions and concepts used in the study reminded, assisted and up skilled the coastal people to rethink about vulnerability, coping mechanisms, adaptive capacity, decision-making powers, access and entitlements to resources, and discriminatory attitudes of society towards women. In this study, the questions and concepts used facilitated the community people to be more public about their problems, coping mechanisms, adaptive capacity, and rights. For example, the lack of access and entitlements fuels risk and vulnerability of women and financial independence, access and entitlement to resources make women more capable in decision-making, which both men and women in the study area did not think about actively before this study.

Natural conversation with the people from different walks of the community helped people in Sutarkhali and Kali Nagor villages to come out of the box of everyday practices, i.e., talk about unequal power relation between men and women, which causes disproportionate vulnerability. Although researcher's presence in the village was for a short time period, but it impacted the community and helped to realise that a longer presence could be more fruitful for future research. Ethnographic research will help people in a community to be more motivated for gender-neutral strategies through additional interaction with the researcher and discussion with each other because of their activities.

Among GOs, NGOs and local people, one potential future research area could be the manner in which coordination, cooperation and network systems work in the coastal area. The results of chaotic coordination between GOs and NGOs bring additional problems in the community, although both are operating in the coastal area. Discussants and NGO activist report that a lack of coordination between GO and NGOs obstructed quality development in the coastal area, which is extremely chaotic during a disaster. It would have been a better idea to examine the lack of coordination between GO and NGOs during the time they work, as well as research could have been performed in disastrous times to shadow activities of GOs-NGOs.

This study primarily focuses on women and a small number of males in the discussion on focus group as well as in-depth individual interviews. In a future study, the views of men could be incorporated in equal numbers to understand the disaster and women's discourse. Perhaps, the incorporation of men and action research would be good to encourage men and women to work together on a project and would improve interaction.

This study was conducted in two different villages from two unions: these unions are geographically located in close proximity to the Bay of Bengal. Both villages are extremely vulnerable to natural hazards and the villagers are from the same socio-cultural background. Although the economic backgrounds of individuals differed in terms of wealth, livelihoods and earning scope were the same such as farming, *gher* owner, shrimp cultivation, and fishing in the river. This study could be replicated on a larger scale in a larger area (i.e., a different area from a different part of the country with disaster, vulnerability and women from different aspects) to compare whether there is any difference in the results. Furthermore, it could be replicated in another coastal area of Bangladesh or anywhere in Bangladesh with the same socio-cultural conditions, livelihood options and earning options or where unequal socio-cultural and economic relations exist to assess the validity of the results of this study.

This study shows that policy-makers, policy planners and practitioners of GO do not pay focus on the voices of the community people; moreover, the voices of the people in disaster-prone areas are ignored. A study could be performed in future to explore the natural constraints while planning and implementing the policies. Further study can address the challenges policy-makers face in involving local community people who are oriented and better experienced about the locality. What are the limitations between grassroots levels and policy levels? Such research may contribute to reducing gender vulnerability and improving adaptive capacity and coping mechanisms.

7.6 Conclusion

This book explored disaster, gender, women, vulnerability and transformation in women's lives, including the role of GOs and NGOs as well as changes in traditional practices in the coastal people of Bangladesh living in extreme poverty. Cyclones are natural hazards, occur very often in Bangladesh and are occurring increasingly often because of global warming and are one of the primary issues affecting people living on the coast. Natural hazards affect women more severely than men. This book focuses on the gender perspective that women are considered disproportionately vulnerable for socio-cultural reasons and because of their limited access to resources in the coastal area. This leads to a stereo typed attitude towards women as helpless victims and interacts with problematic cultural practices, customs and traditions that ignore the contribution of women to the family's economy and community. Women's coping and adaptation mechanisms are undermined; therefore, their ability to use traditional knowledge to cope in disastrous times is affected. All these limitations are examined alongside the nature of women's household responsibilities and restriction on women's migration thus creating double responsibilities on the shoulders of women.

Disaster studies-related concepts such as the *Pressure and Release Approach and Access and Entitlements to Resources* models provide a background to the theoretical framework for assessing the vulnerability of women in the coastal area of Bangladesh in the context of limited resources and income. In this study, *Feminist Political Ecology* and Connell's *Hegemonic Masculinity* have been incorporated with the aim to understand gender relationships in the context of the local environment and natural resources. *Theories of Social Change* complement the analysis of policy-makers and policy practitioners who actively consider additional participation and additional involvement of women with development policy. As per the mixed methods approach, this in-depth field research has been undertaken in two different disaster-prone villages in the coastal area in Bangladesh. Supplemented by nine focus groups and individual interviews, a survey of 263 women collected data on.

- the impact of unequal distribution and lack of access/entitlements to resources
- institutional local practices causing disproportionate vulnerability for women
- participation of women in the decision-making process within the family and the community
- coping and adaptive mechanisms taken by women and female-headed households in disastrous times
- problems in promoting women's participation in their communities and leadership
- changes in attitudes and local practices towards women

This study assesses the discourse of disaster, gender and women in the coastal area examining the policies and role of GOs and NGOs, contrasting them with what is actually happening in the life of coastal people. GOs are providing work programmes and disaster preparedness, management and relief programmes that target women's requirements. NGOs and women's organisations are placing their central efforts into changing ordinary people's attitudes towards gender-related roles. People in this study argued that natural forces such as climatic variability have triggered responses to these programmes that have started to change the existing established and long-practised gender relations. Finally, this book argues that gender equality, recognition of women's contribution, ensuring participation in decision-making, space for leadership, and freedom of mobility beyond their households will ensure a better approach of life for coastal people.

Glossary of Terms

Aila Name of a cyclone that hit the study area in 2009

Apa Sister

Bangladesh Mohila Parishad Bangladesh Council for Women

Bari House

Bazar Place for buying and selling goods in the village is held once or twice in a week

Beribadh Dams on the riverbank

Bhabi Sister-in-law

bKash Money-sending system via mobile phones and mobile banking in Bangladesh; the client does not require to go to the bank

Bokhate Unruly/immoral boys of locality

Borkha Veil over garments that cover the whole body of women to hide their appearance and any part of body from being seen by outsiders

Boudi Sister-in-law

Chachi Aunt

Char Island in river separated/detached from the mainland

Dadi Grandmother

Dadon Informal cash loan with a high interest rate

Deribadha Day-time labourer

Didi Sister

Dima An aged women who assists with a baby's birth in the village

Dormiyo bisshas Religious beliefs

Durjog Disaster

Ghar bhanda Ensuring a residence free from evils entering by hanging a citation from the Quran at the front and back door of the house given by Kabiraj and a religious leader

Gher Type of big pond/lake artificially created for cultivating fish, primarily shrimps

Gholpata A type of long grass/leaves available in *Sundarban* used for making houses

Hudhud The name of a cyclone that happened in the study area in 2014

Jamela Problems/dispute

M. M. Khan, *Disaster and Gender in Coastal Bangladesh*,
https://doi.org/10.1007/978-981-19-3284-7

Jhupri Houses/residences composed of leaves and grass

Jungling Collecting honey, wood, crab and fish from Sundarbans (mangrove forests area in the coastal area in Bangladesh)

Kabikha Food for work

Kabiraj A man providing traditional herbal and religious knowledge-based treatment in the village

Kajer Binimoye Khaddo Food for work

Katrik A Bangla month identified by the local people as a month of scarcity

Khashjami Abandoned and unused land owned by the government

Lakh One hundred thousand

Lok dekhano Eye wash

Madrasha A religion-based educational institution

Mahila Samity Informal cooperatives/associations formed by the local women

Mattabbor Local village leader

Maya Affection/belongingness

Mohajon Man who gives/provides informal cash loans in villages having a high interest rate

Mashi Aunt

Mati Kata Developing dams/roads and digging canals

Matirkaj Day labourer

Miking Disseminating information/warning in the locality via loud speakers or microphones

Mor A common place for villagers to gather during leisure/free time

Nani Grandmother

Nargis Name of a cyclone that happened in 2008

Nari Bikash Kendra Development and coordination centre for women

Naripokko An NGO that works for women

Niriho Distressed and helpless

Pacca Brick/concrete built

Para Community

Paribarik jamela Household-related disputes/problems

Parishad Council

Purdha Curtain or veil that covers the body of women, hiding their appearance and any part of the body from being seen by outsiders and keeping them confined in the house

Sadar Headquarters/office of union and upazila

Samity Informal cooperatives/associations formed by local people

Sangstha Organisation primarily local people mean NGOs

Saree A piece of unstitched 10/12 feet long cloth used by women

Shangathan Organisation

Shalish A traditional alternative dispute resolution in the rural area conducted by local elite and influential people

Shaola bacha Cleaning grass and parasite from *gher* as day labourer

Shariah Islamic law based on the teachings of Quran and traditions of the Prophet (Hadith and Sunna)

Shoye jaowa Shock absorber/get used to

Sidr Name of a cyclone that happened in the coastal area in 2007

Sundarban The largest mangrove forest in the world in the study area. It ranges from Bangladesh's Khulna Division to West Bengal in India.

Tabij Amulet given by Kabiraj as treatment for various diseases to hang around necks or arms

Thana Police station

Trawler A motor boat is locally known as a trawler

Union The smallest rural administrative and local government units in Bangladesh. Each union is composed of nine wards; usually, one village is designated as a ward.

Union shastha kendra Union Health Centre/Complex located at Union Parishad Headquarters

Unnayan onneshan Searching/looking for development

Upazila Sub-district—A third tier of local government in Bangladesh

Wabda Dam/polder on the river bank to protect the community from surge or tidal water known as wabda by the local people

References

Adger, W. N. (2006). Vulnerability. *Global Environmental Change, 16*(3), 268–281.

Adger, W., Neil, P. Y., Huq, S., & Mace, M. J. (2006). *Fairness in adaptation to climate change.* MIT Press.

Adger, W. N., Agrawala, S., Mirza, M. M. Q., Conde, C., O'Brien, K., Pulhin, J., Pulwarty, R., … Takahashi, K. (2007). *Assessment of adaptation practices, options, constraints and capacity* (IPCC's Fourth Assessment Report, pp. 719–743).

Adu-Poku, S. (2001). Envisioning (Black) male feminism: A cross-cultural perspective. *Journal of Gender Studies, 10*, 157–167.

African Development Bank Group. (2009). *Checklist for gender mainstreaming in the infrastructure sector.* Retrieved from http://www.afdb.org/fileadmin/uploads/afdb on June 10, 2016.

Afrin, S., Islam, N., & Ahmad, S. U. (2008). A multivariate model of micro credit and rural women entrepreneurship development in Bangladesh. *International Journal of Business and Management, 3*(8), 169–185.

Agarwal, B. (1986). Women, poverty and agricultural growth in India. *Journal of Peasant Studies, 13*(4), 165–200.

Agarwal, B. (1992). The gender and environment debate: Lessons from India. *Feminist Studies, 18*(1), 119–158.

Agarwal, B. (1997). 'Bargaining' and gender relations: Within and beyond the households. *Feminist Studies, 3*(1), 1–51.

Agarwal, B. (2001). Participatory exclusions, community forestry, and gender: An analysis for South Asia and a conceptual framework. *World Development, 29*(10), 1623–1648.

Ahmed, S. M., Adams, A., Chowdhury, A. M. R., & Bhuiya, A. (1997). Income-earning women from rural Bangladesh: Changes in attitudes and knowledge, empowerment. *Women for Women, 4*, 1–12.

Ahmed, A. U., Neelormi, S., & Adri, N. (2007). *Entrapped in water world: Impacts and adaptation to climate change induced water logging for women in Bangladesh.* Centre for Global Change.

Ahmed, A. U., Neelormi, S., Adri, N., Alam, S., & Nuruzzaman, K. (2008). *Climate change, gender and vulnerable groups in Bangladesh.* Climate Change Cell, MoFDM.

Ahmed, A. U., Quisumbing, A. R., Nasreen, M., Hoddinott, J. F., & Bryan, E. (2009). *Comparing food and cash transfers to the ultra poor in Bangladesh.* Research Monograph 163: International Food Research Institute.

Ahmed, A. U., Haq, S., Nasreen, M., & Hassan, A. W. R. (2015). *Climate change and disaster management: Sectoral inputs towards the formulation of Seventh Five Year Plan (2016–2021).* Planning Commission.

M. M. Khan, *Disaster and Gender in Coastal Bangladesh*, https://doi.org/10.1007/978-981-19-3284-7

Akhter, F. (1992). Women are not only victims. In H. Hameeda, D. Cole, & F. H. Abel (Eds.), *From crisis to development: Coping with disasters in Bangladesh* (pp. 59–66). University Press.
Alam, K., & Rahman, M. H. (2014). Women in natural disasters: A case study from southern coastal region of Bangladesh. *International Journal of Disaster Risk Reduction, 8*, 68–82.
Alston, M. (2015). *Women and climate change in Bangladesh.* Routledge.
Ariyabandu, M. (2003). Voice of women. *Journal of Women's Liberation, 6*(1), 49–59.
Ariyabandu, M. M. (2009). Sex, gender and gender relations in disasters. In E. Enarson & P. G. Dhar Chakrabarti (Eds.), *Women, gender and disaster: Global issues and initiatives* (pp. 5–17). Sage.
Asian Development Bank [ADB]. (2004). *Country environmental analysis.*
Asian Development Bank [ADB]. (2010). *The informal sector and informal employment in Bangladesh: Country Report 2010.* Asian Development Bank.
Ayers, J., Huq, S., Wrught, H., Faisale, A. M., & Hussain, S. T. (2014). Mainstreaming climate change adaptation into development in Bangladesh. *Climate and Development, 6*(4), 293–305.
Azad, A. K. (2013, September). *Climate change and livelihood in Bangladesh: Experiences of people living in coastal regions.* Paper presented at the International Conference on Building Resilience organized by the Centre for Disaster Resilience, University of Salfor, Ahungalla, Sri Lanka.
Bachelet, M. (2012). *Rio+20 sustainability summit.* Retrieved from www.unwomen.org on February 23, 2018.
Backer, T. E. (2001). Increasing participation means changing behavior: What can be learned from behavioral science? *Grantmakers in the Arts Reader, 12*(1), 18–22.
Baden, S., Green, C., Goetz, A. M., & Guhathakurta, M. (1994). *Background reports on gender issues in Bangladesh* (BRIDGE Reports 26). Institute of Development Studies.
Banford, A., & Froude, C. K. (2015). Ecofeminism and natural disasters: Sri Lankan women post-tsunami. *Journal of International Women's Studies, 16*, 170–187.
Bari, S. (1992). Women in the aftermath. In H. Hameeda, D. Cole, & F. H. Abel (Eds.), *From crisis to development: Coping with disasters in Bangladesh* (pp. 55–58). University Press.
Basak, S. R., Basak, A. C.,& Rahman, M. A. (2017). *Women empowerment to lead change-Bangladesh perspective.* Retrieved from http://feppcar.org/396/women-empowerment-to-lead-change-bangladesh-perspective/ on June 26, 2019.
Baten, A. M., & Khan, A. N. (2010). *Gender issue in climate change discourse: Theory versus reality.* UnnayanOnneshan.
Bates, D. C. (2002). Environmental refugees? Classifying human migrations caused by environmental change. *Population and Environment, 23*, 465–477.
BBS [Bangladesh Bureau of Statistics]. (2008). *Gender statistics of Bangladesh.* Statistics and Information Division. Ministry of Planning.
BBS [Bangladesh Bureau of Statistics]. (2010). *Bangladesh household income and expenditure survey.* Statistics and Information Division.
BBS [Bangladesh Bureau of Statistics]. (2011). *Population and housing census.* Statistics Division and Information Division.
BBS [Bangladesh Bureau of Statistics]. (2012). *Gender statistics of Bangladesh.* Statistics and Information Division.
BBS [Bangladesh Bureau of Statistics]. (2018). *Gender statistics of Bangladesh.* Statistics and Information Division.
Bebbington, A. (2005). Donor–NGO relations and representations of livelihood in non-governmental aid chains. *Journal of World Development, 33*(6), 937–950.
Beck-Gernsheim, E., Butler, J., & Puigbert, L. (2003). *Women and social transformation.* Peter Lang.
Begum, S., & Sen, B. (2009). Maternal health, child well-being and chronic poverty: Does women's agency matter? *The Bangladesh Development Studies, XXXII*(4), 69–93.
Bertelsmann Foundation. (2012). *Bangladesh Country Report.*

Bhandari, R. B. (2014). Social capital in disaster risk management; a case study of social capital mobilization following the 1934 Kathmandu Valley earthquake in Nepal. *Disaster Prevention and Management, 23*(4), 314–328.

Birkmann, J., Cardona, O. D., Carreño, M. L., Barbat, A. H., Pelling, M., Schneiderbauer, S., … Welle, T. (2013). Framing vulnerability, risk and societal responses: the MOVE framework. *Natural Hazards, 67*(2), 193–211.

Birkmann, J., & Fernando, N. (2008). Measuring revealed and emergent vulnerabilities of coastal communities to tsunami in Sri Lanka. *Disasters, 32*(1), 82–105.

Blaikie, P., Cannon, T., Davis, I., & Wisner, B. (1994). *At risk: Natural hazards, people's vulnerability, and disaster.* Routledge.

Blalock, H. M. (1985). *Social statistics.* McGraw-Hill.

Bolin, R., Martina, J., & Allison, C. (1998). Gender inequality, vulnerability, and disaster: Issues in theory and research. In E. Enarson & B. H. Morrow (Eds.), *The gendered terrain of disaster: Through women's eyes* (pp. 27–44). Praeger.

Bose, M. L., Ahmad, A., & Hossain, M. (2009). The role of gender in economic activities with special reference to women's participation and empowerment in rural Bangladesh. *Gender Technology and Development, 13*(1), 69–102.

Bradshaw, S. (2001). Reconstructing roles and relations: Women's participation in reconstruction in post-Mitch Nicaragua. *Gender and Development, 9*(3), 79–87.

Bradshaw, S. (2013). *Gender, development and disasters.* Edward Elgar.

Bradshaw, S. (2014). Engendering development and disasters. *Disasters, 39*(S1), 54–75.

Bradshaw, S., & Fordham, M. (2014). Double disaster: Disaster through a gender lens. In F. S. John, E. C. Andrew, E. C. Andrew, M. Bernard, & J. Janaka (Eds.), *Hazards, risks and disasters in society* (pp. 233–252). Elsevier.

Bradshaw, S., Chant, S., & Linneker, B. (2017). Gender and poverty: What we know, don't know, and need to know for Agenda 2030. *Gender, Place and Culture, 24*(12), 1667–1688.

Brakenridge, G. R., Syvitski, J. P. M., Overeem, I., Higgins, S. A., Kettner, A. J., Stewart-Moore, J. A., & Westerhoff, R. (2012). Global mapping of storm surges and the assessment of coastal vulnerability. *Natural Hazards, 66*(3), 1295–1312.

Bryan, G., Chowdhury, S., & Mushfiq, A. (2014). Underinvestment in a profitable technology: The case of seasonal migration in Bangladesh. *Econometrica, 82*(5), 1671–1748.

Bryman, A. (2012). *Social research methods.* Oxford University Press.

Butler, J. (2004). Introduction: Acting in concert. In J. Butler (Ed.), *Undoing gender* (pp. 1–16). Routledge.

Cannon, T. (2002). Gender and climate hazards in Bangladesh. *Gender and Development, 10*(2), 45–50.

Cannon, T. (2008). Vulnerability, "innocent" disasters and the imperative of cultural understanding. *Disaster Prevention and Management, 17*(3), 350–357.

Carney, D. (2002). *Sustainable livelihoods approaches: Progress and possibilities for change.* Department for International Development.

Chadwick, B. (2002). *Beyond conflict to consensus: An introductory learning manual.* Retrieved from www.managingwholes.com/chadwick.htm on April 7, 2015.

Chafetz, J. S. (1997). Feminism theory and sociology: Underutilized contributions for mainstream theory. *Annual Review of Sociology, 23*, 97–120.

Chant, S. (2008). Dangerous equations? How female headed households become poorest of the poor: Causes, consequences and cautions. In J. D. Momsen (Ed.), *Gender and development: Critical concepts in development studies* (pp. 397–409). Routledge.

Chen, M. A. (1986). Poverty, gender and work in Bangladesh. *Economic and Political Weekly, XXi*(5), 217–222.

Chen, L. C., Huq, E., & D'Souza, S. (1981). Sex bias in the family allocation of food in rural Bangladesh. *Population and Development Review, 7*(1), 55–70.

Chowdhury, A. K. M., Huffman, S. L., & Chen, L. C. (1981). Agriculture and nutrition in Matlab Thana, Bangladesh. In R. Chamber, R. Longhurst, & A. Pacey (Eds.), *Seasonal dimensions of rural poverty* (pp. 52–61). Frances Pinter.

Chowdhury, M. (2001, November). *Women's technological innovations and adaptations for disaster mitigation: A case study of charlands in Bangladesh* prepared for DAW/ISDR Expert Group Meeting on "Environmental Management and the Mitigation of Natural Disasters: A Gender Perspective." Ankara, Turkey.

Chowdhury, M., F. I. (2015, August 5). Revisiting position of women in Bangladesh. *The Daily Star*.

Connell, R. W. (1995). *Masculinities*. Polity Press.

Connell, R. W. (1996). Teaching the boys: New research on masculinity and gender strategies for schools. *Teachers College Record, 98*, 206–235.

Connell, R. W. (2005). Hegemonic masculinity: Rethinking the concept. *Gender and Society, 19*(6), 829–859.

Connell, R. W. (2006). Masculinities and globalization. In E. Stanley & M. Baca Zinn (Eds.), *Globalization: The transformation of social world* (pp. 216–230). Thomson Wadsworth.

Corbin, L., & Strauss, A. (2008). *Basics of qualitative research: Techniques and procedures for developing grounded theory*. Sage.

CPD [Center for Policy Dialogue]. (2007).*Bangladesh vision 2021*. Nagorik Committee.

CRED [Centre for Research on the Epidemiology of Disasters]. (2015). *The Human cost of natural disasters 2015: A global perspective*. CRED.

Creswell, J. W. (2003). *Research design: Qualitative, quantitative and mixed methods approach*. Sage.

Creswell, J. W. (2007). *Designing and conducting mixed methods research*. Sage.

Creswell, J. W., Clark, P., Gutmann, M., & Hanson, W. (2003). Advanced mixed methods research designs. In A. Tashakkori & C. Teddlie (Eds.), *Handbook for mixed methods in social and behavioural research* (pp. 161–196). Thousand Oaks.

Crook, R. C., & Manor, J. (1998). *Democracy and decentralization in South-East Asia and West Africa: Participation, accountability and performance*. Cambridge University Press.

Cunha, J. D. (1997). Engendering disaster preparedness and management. *Asian Disaster Management News, 3*(3), 2–5.

Cutter, S. L., Boruff, B. J., & Shirley, W. L. (2003). Social vulnerability to environmental hazards. *Social Science Quarterly, 84*, 242–261.

Dankelman, I. (2002). Climate change: Learning from gender analysis and women's experiences of organizing for sustainable development. In R. Masika (Ed.) *Gender, development and climate change* (pp. 21–29). Oxfam.

Dankelman, I. (2010). Introduction: Exploring gender, environment and climate change. In I. Dankelman (Ed.), *Gender and climate change: An introduction* (pp. 1–20). Routledge.

Daly, K. (1997). Different ways of conceptualizing sex/gender in feminist theory and their implications for criminology. *Theoretical Criminology, 1*, 25–51.

Dalrymple, S., Hiscock, D., Azad, A. B., Husain, N., & Rahman, A. B. M. Z. (2009). *Climate change and security of Bangladesh: A case study*. Bangladesh Institute of International and Strategic Studies (BIISS) and Safer World.

Dasgupta, S. (2010). Introduction. In S. Dasgupta, I. Siriner, & P. Sarathi De (Eds.), *Women's encounter with disaster* (pp. 1–19). Front Page Publications.

Dasgupta, S., Siriner, I., & Sarathi, P. D. (Eds.). (2010). *Women's encounter with disaster*. Front Page Publications.

Dasgupta, S., Yeates, N., Driskell, R., & Braun, Y. A. (Eds.). (2012). *Women's encounter with globalization*. FrontPage Publications.

Delica, Z. G. (1998). Balancing vulnerability and capacity: Women and children in the Philippines. In E. Enarson & B. H. Morrow (Eds.), *The gendered terrain of disaster: Through women's eye* (pp. 109–113). Praeger Publishers.

Demetriades, J., & Esplen, E. (2008). The gender dimensions of poverty and climate change adaptation. *IDS Bulletin, 29*(4), 24–31.

Denissen, A. K. (2012). *Climate change and its impacts on Bangladesh.* Retrieved from www.ncd o.nl, on 17 January 2019.

Denton, F. (2000). Gender and climate change: Giving the "latecomer" a head start". *IDS Bulletin, 35*(3), 42–49.

Denton, F. (2002). Climate change vulnerability, impacts and adaptation: Why does gender matter? *Gender and Development, 10*(2), 10–20.

Devereux, S. (2001). Sen's entitlement approach: Critiques and counter-critiques. *Oxford Development Studies, 29*(3), 245–263.

Diamond, I., & Orenstein, G. F. (1990). *Reweaving the world: The emergence of ecofeminism.* Sierra Club Books.

Dobson, N. (1994). From under the mud-pack: Women and Charlesville floods. *Australian Journal of Emergency Management, 9*(2), 11–13.

Dreze, J., & Sen, A. (1989). *Hunger and public action.* Oxford University Press.

Enarson, E. (2000). *Gender and natural disasters.* Working paper 1, in the programme focusing on 'Crisis Response and Reconstruction'. Recovery and Reconstruction Department, IOM.

Enarson, E. (2012). *Women confronting natural disaster from vulnerability to resilience.* Lynne Rienner.

Enarson, E., & Meyreles, L. (2004). International perspectives on gender and disaster: Differences and possibilities. *The International Journal of Sociology and Social Policy, 24*(10/11), 49–93.

Enarson, E., & Morrow, B. H. (1998a). *The gendered terrain of disaster: Through women's eye.* Praeger.

Enarson, E., & Morrow, B. H. (1998b). Why gender? Why women? An introduction to women and disaster. In E. Enarson & B. H. Morrow (Eds.), *The gendered terrain of disaster: Through women's eye* (pp. 1–8). Praeger.

Enarson, E., & Pease, B. (2016). The gendered terrain of disaster: Thinking about men and masculinities. In E. Enarson & B. Pease (Eds.), *Men, masculinities and disaster* (pp. 3–19). Routledge.

Faisal, I. M., & Kabir, M. R. (2005). An analysis of gender-water nexus in rural Bangladesh. *Journal of Developing Societies, 21*(1–2), 175–194.

Fakir, S. K. (2008). *Women empowerment through participation in income generating activities of Sabalamby Unnayan Samity* (Ph.D. thesis). Department of Agricultural Extension Education, Bangladesh Agricultural University.

Farhana, K., Sayduzzaman, M., & Munir, M. S. (2015). Present status of workers in readymade garments industries in Bangladesh. *European Scientific Journal, 11*(7), 45–53.

Ferdousi, F. (2015). Factors affecting implementation of new public management in Bangladesh. *International Journal of Management Sciences., 5*(5), 315–328.

Fernando, P., & Fernado, V. (1997). *South Asian women: Facing disaster, securing life.* Duryog Nivaran and Intermediate Technology Group.

Feldman, S. (2001). Exploring theories of patriarchy: A perspective from contemporary Bangladesh. *Signs: Journal of Women and in Culture and Society, 26*(4), 1097–1127.

Fisher, S. (2010). Violence against women and natural disasters: Findings from post-tsunami Sri Lanka. *Violence against Women, 16*, 902–918.

Fiske, S. T., & Berdahl, J. (2007). Social power. In A. W. Kruglanski & E. T. Higgins (Eds.), *Social psychology: Handbook of basic principles* (pp. 678–849). Guilford.

Fordham, M., & Ketteridge, A. M. (1998). Men must work and women must weep: Examining gender stereotypes. In E. Enarson & B. H. Morrow (Eds.), *The gendered terrain of disaster: Through women's eye* (pp. 1–8). Praeger.

Fothergill, A. (1998). The neglect of gender in disaster work: An overview of the literature. In E. Enarson & B. H. Morrow (Eds.), *The gendered terrain of disaster: Through women's eye* (pp. 11–25). Praeger.

Fothergill, A. (2004). *Heads above water: Gender, class, and family in the Grand Forks Flood*. University of New York Press.

Fothergill, A., & Peek, L. (2004). Poverty and disasters in the United States: A review of recent sociological findings. *Natural Hazards, 32*, 89–110.

Frith, H. (2000). Focusing on sex: Using focus groups in sex research. *Sexualities, 3*, 275–297.

Fuchs, S., Kuhlicke, C., & Meyer, V. (2011). Vulnerability to natural hazards-the challenge of integration. *Natural Hazards, 58*(2), 609–619.

Fukuyama, F. (2001). Social capital, civil society and development. *Third World Quarterly, 22*(1), 7–20.

Fussel, H. M. (2007). Vulnerability: A generally applicable conceptual framework for climate change research. *Global Environmental Change, 17*(2), 155–167.

Gaillard, J., Fordham, M., & Sanz, K. (2015). Culture and disasters: Understanding cultural framings in disaster risk reduction In F. Krüger, G. Bankoff, T. Cannon, B. Orlowski, & E. L. F. Schipper (Eds.), *Culture and disasters: Understanding cultural framings in disaster risk reduction* (pp. 224–234). Taylor & Francis.

Ginige, K., Amarratunga, D., & Haigh, R. (2015). *Mainstreaming women into disaster risk reduction in the built environment in Sri Lanka*. Retrieved from www.pure.hud.ac.uk on December 10, 2019.

Giddens, A. (2009). *Sociology*. Polity Press.

GoB [Government of the People's Republic of Bangladesh]. (2008). *Report of the Labour Force Survey*. Ministry of Planning.

GoB [Government of the People's Republic of Bangladesh]. (2009a). *Bangladesh Climate Change Strategy and Action Plan (BCCSAP)*. Ministry of Environment and Forests.

GoB [Government of the People's Republic of Bangladesh]. (2009b). *National Adaptation Programme of Action (NAPA)*. Ministry of Environment and Forests.

GoB [Government of the People's Republic of Bangladesh]. (2010a). *Standing order on disaster*. Disaster Management Bureau.

GoB [Government of the People's Republic of Bangladesh]. (2010b). *National plan for disaster management 2010–2015*. Disaster Management Bureau.

GoB [Government of the People's Republic of Bangladesh]. (2011a). *The 6th Five Year Plan FY2011a–FY2015. Accelerating growth and reducing poverty*. Planning Commission, Ministry of Planning.

GoB [Government of the People's Republic of Bangladesh]. (2011b). *National Women Development Policy*. Ministry of Women and Children Affairs (MoWCA).

GoB [Government of the People's Republic of Bangladesh]. (2012). *National Women Development Policy-2011*. Ministry Women and Child Affairs (MoWCA).

GoB [Government of the People's Republic of Bangladesh]. (2013). *Bangladesh climate change and gender action plan (ccGAP: Bangladesh)*. Ministry of Environment and Forests.

GoB [Government of the People's Republic of Bangladesh]. (2016). *The 7th Five Year Plan FY2016–FY2020: Accelerating growth, empowering citizens*. General Economic Division (GED), Planning Commission.

Gokhale, V. (2008). *Role of women in disaster management: An analytical study with reference to Indian society*. Paper presented at the 14th World Conference on Earthquake Engineering, Beijing, China.

Groenhout, R. E. (2002). Essentialist challenges to liberal feminism. *Journal of Social Practice and Theory, 28*(1), 51–75.

Gruen, L. (1993). Dismantling oppression: An analysis of the connection between women and animals. In G. Gaard (Ed.), *Ecofeminism: Women, animals, nature* (pp. 60–90). Temple University Press.

Haider, R., & Akhter, R. (1999). The role of NGO and women's perception of empowerment: An anthropological study in a village. *Women for Women, 6*, 58–62.

Haq, Z., Robbani, M., Ali, M., Hasan, M., Hasan, M., Uddin, J., & Karim, R. (2012). Damage and management of cyclone Sidr-affected homestead tree plantations: A case study from Patuakhali, Bangladesh. *Natural Hazards, 64*, 1305–1322.

Haque, S. (2018). *Women falling behind men in recent BCS exams*. bdnews24.com. Retrieved https://bdnews24.com/bangladesh/2018/04/09/women-fall-behind-men-in-bangladesh-civil-service-bcs-exams on January 22, 2019

Haque, S. K. T. M., & Ali, K. S. (2011). NGO governance: Accountability and transparency issues for the development sector. *Public Management Review, 5*(2), 267–278.

Harvard Humanitarian Initiative [HHI]. (2010). *Hardship of women's mobility and disaster vulnerability*. Harvard University.

Hoque, M., & Itohara, Y. (2008). Participation and decision making role of rural women in economic activities: A comparative study for members and non-members of the micro-credit organizations in Bangladesh. *Journal of Social Sciences, 4*(3), 229–236.

Horton, L. (2012). After the earthquake: Gender inequality and transformation in post disaster Haiti. *Gender and Development, 20*(2), 295–308.

Hossain, M., Bose, M. L., & Ahmad, A. (2004). *Nature and impact of women's participation in economic activities in rural Bangladesh: Insights from household surveys*. Lund University.

Hovorka, A. J. (2006). The no.1 ladies' poultry farm: A feminist political ecology of urban agriculture in Botswana. *Gender, Place and Culture, 13*(3), 207–225.

HSBC. (2011). *Climate investment update*. HSBC Global Research.

Huq, S., & Karim, Z. (1999). *Vulnerability and adaptation to climate change for Bangladesh*. Kluwer Academic.

Huq, S., Reid, H., Konate, M., Rahman, A., Sokona, Y., & Crick, F. (2004). Mainstreaming adaptation to climate change in least developed countries (LDCs). *Climate Policy, 4*(1), 25–43.

IFRC [International Federation of Red Cross and Red Crescent Societies]. (2010). *Bangladesh: Cyclone Aila*.

Iftekharuzzaman. (2011). Corruption and anti-corruption in Bangladesh: Primacy of the political. Presented at the seminar on *"Bangladesh at 40: Changes and Challenges"*, organized by the Faculty of Business Studies, Jahangirnagar University, 9–11 December 2011.

Ikeda, K. (1995). Gender differences in human loss and vulnerability in natural disasters: A case study from Bangladesh. *Indian Journal of Gender Studies, 2*(2), 171–193.

Ikeda, S. (2006). An integrated risk analysis framework for emerging disaster risks: Toward a better risk management of flood disaster in urban communities. In S. Ikeda, T. Fukuzono, & T. Sato (Eds.), *A better integrated management of disaster risks: Toward resilient society to emerging disaster risks in megacities*. Terra Scientific Publishing.

IPCC [Inter-governmental Panel on Climate Change]. (2007). *Climate Change 2007: Synthesis Report*. Retrieved from http://www.ipcc.ch/pdf/assessment-report/ar4/syr/ar4_syr.pdf on 22 November 2013.

Islam, N. (2015). *Corruption, its control and drivers of change: The case of Bangladesh*. Bangladesh Institute of Development Studies.

Jahan, M. (2008). The impact of environmental degradation on women in Bangladesh: An overview. *Asian Affairs, 30*(2), 5–15.

Jahan, M. (2007). Gender mainstreaming in Bangladesh civil service: Prospects and constraints. *Asian Affairs, 29*(1), 41–72.

Jain, R. (2017). Conceptualizing empowerment. In R. Jain (Ed.), *Women empowerment: The global scenario* (pp. 239–240). Studium Press (India).

Jim, W. M. H., & Hossain, M. (2011). *Women's participation in agriculture in Bangladesh 1988–2008: Changes and determinants*. Paper presented in the pre-conference event on 'Dynamics of rural livelihoods and poverty in South Asia'. The 7th Asian Society of Agricultural Economists (ASAE). International Conference Hanoi, Vietnam.

Jinnah, S. I. M. (2013). *Land and property rights of rural women in Bangladesh*. Community Development Association.

Johnston, M. (2017). *Good governance: Rule of law, transparency and accountability*. Colgate University.

Kabeer, N. (1990). *Minus lives: Women of Bangladesh*. CHANGE International Reports, Women and Society.

Kabeer, N. (1999). Resources, agency, achievements: Reflections on the measurement of women's empowerment. *Development and Change, 30*, 35–464.
Kabeer, N. (2001a). Reflection on the measurement of women's empowerment. In *Discussing Women's empowerment: Theory and practice.* Novum Grafiska.
Kabeer, N. (2001b). Conflicts over credit: Re-evaluating the empowerment potential of loans to women in rural Bangladesh. *World Development, 29*(1), 63–84.
Kabeer, N. (2011). Between affiliation and autonomy: Navigating pathways of women's empowerment and gender justice in rural Bangladesh. *Development and Change, 42*(2), 499–528.
Kabeer, N., Mahmud, S., & Castro, J. G. I. (2012). NGOs and the political empowerment of poor people in rural Bangladesh: Cultivating the habits of democracy? *World Development, 40*(10), 2044–2062.
Kabeer, N., Mahmud, S., & Castro, J. G. I. (2010a). *NGOs' strategies and the challenge of development and democracy in Bangladesh* (IDS Working Paper 343).
Kabir, K. (1992). How women survived. In H. Hossain, C. P. Dodge, & F. H. Abed (Eds.), *From crisis to development: Coping with disasters in Bangladesh* (pp. 71–81). University Press.
Karlsson, G. (2007). *Energy as women's business.* Retrieved from http://www.energia.org/cms/wp-content/uploads/2015/06/56-Energy-as-womens-business.pdf on 11 January 2016.
Khan, A. H. (1995). *Ganges water sharing: An overview in women for water sharing.* Academic.
Khan, M. I. (2013). Social changes in contemporary Bangladesh. *Journal of the Asiatic Society of Bangladesh (Hum.), 58*(2), 263–276.
Khan, M. M., & Azad, A. K. (2012). The nexus between local politics and salinity intrusion in coastal area of Bangladesh: Unveiling the development myth. *The Journal of Social Development, 24*(1), 65–76.
Khaled, S. S. (2018). Scale of women empowerment in Bangladesh. *The Daily Independent.* http://www.theindependentbd.com/post/165767. Retrieved on December 2, 2019.
Khatoon, K. (1995). *Impact of Shrimp cultivation: Pattern of changes in the lives of people in Paikgachha and Shyamnagar* (Unpublished research report). Ain-O-Shalish Kendra.
Khondker, H. H. (1996). Women and floods in Bangladesh. *International Journal of Mass Emergencies and Disasters, 14*(3), 281–292.
Khondker, H. H. (2014). *Social transformation, development and globalization.* Retrieved from sociopedia.isa on May 26, 2019.
Kimmel, M. S. (2008). *The gendered society.* Oxford University Press.
Kishor, S., & Gupta, K. (2004). Women's empowerment in India and its states: Evidence from the NFHS. *Economic and Political Weekly, 39*(7), 694–712.
Knox, C. (2009). Dealing with sectoral corruption in Bangladesh: Developing citizen involvement. *Public Administration and Development, 29*, 117–132.
Kochanek, S. A. (1996). Bangladesh in 1996: The 25th year of independence. *Asian Survey, 37*(2), 136–142.
Kron, W. (2012). Coasts: The high-risk areas of the world. *Natural Hazards, 66*(3), 1363–1382.
Kumar-Range, S. (2001). *Environmental management and disaster risk reduction: A gender perspective.* Prepared for DAW/ISDR expert group meeting on 'Environmental Management and the Mitigation of Natural Disasters: A Gender Perspective' in Ankara, Turkey.
Leat, D. (2005). *Theories of social change.* Bertelsmann Stiftung: International Network on Strategic Philanthropy (INSP). Paper 4.
Lee, D. (1997). Interviewing men: Vulnerabilities and dilemmas. *Women's Studies International Forum, 20*(4), 553–564.
Leighton, M. S., & Wraner, K. (Eds.). (2011). *Climate change and migration: Rethinking policies for adaptation and disaster risk reduction.* Publication Series of UNU-EHS, 15/2011. Studies of the University, Research, Counsel, Education.
Longenecker, M. (1997). Women, ecology, and the environment: An introduction. *National Women's Studies Association Journal, 9*(3), 1–17.
MacGregor, S. (2010). A stranger silence still: The need for feminist social research on climate change. *The Sociological Review, 57*, 124–140.

Mahmud, K. T., Islam, S., Parvez, A., & Haque, S. (2017). Impact of fishery training programs on the household income of the rural poor women in Bangladesh. *Human Resource Management Research, 7*(3), 97–101.

Mahmud, K. T., Hossain, S., Parvez, A. & Hilton, D. (2017b). Agricultural risk management capacity of the rural women borrowers under micro credit programme of BRAC: Bangladesh perspective. In *Women empowerment: The global scenario* (pp. 241–153). Stadium Press (India).

Mahmud, K. T., Parvez, A., Hilton, D., Kabir, G. M. S., & Wahid, I. S. (2014). The role of training in reducing poverty: The case of agricultural workers receiving microcredit in Bangladesh. *International Journal of Training and Development, 18*(4), 283–290.

Mahmud, K. T. (2010). Does microcredit really matter to women borrowers of Bangladesh? A case of BRAC's agribusiness program. *The Global Journal of Finance and Economics, 7*(2), 111–121.

Mallick. B., & Vogt, J. (2009). Analysis of disaster vulnerability for sustainable coastal zone management: A case of cyclone *Sidr* 2007 in Bangladesh. In *Climate change: Global risks, challenges and decisions*. Conference Series, Earth and Environmental Science 6.

Mallick, B., Rahaman, K. R., & Vogt, J. (2011). Coastal livelihood and physical infrastructure in Bangladesh after cyclone *Aila. Mitigation Adaptation Strategies Global Change, 16*(6), 629–648.

Mallick, B., & Vogt, J. (2012). Cyclone, coastal society and migration: Empirical evidence from Bangladesh. *International Development Plan Review, 34*(3), 217–240.

Mallick, B., & Vogt, J. (2014). Population displacement after cyclone and its consequences: Empirical evidence from coastal Bangladesh. *Natural Hazards, 73*(2), 191–212.

Manju, T. (2006). *Political culture of shrimp culture on Bangladesh* (Poverty Research Report 13). Grameen Trust.

Manyena, S. B. (2012). Disaster and development paradigms: Too close for comfort. *Development Policy Review, 30*, 327–345.

Marilley, S. M. (1996). *Woman suffrage and the origins of liberal feminism in the United States 1820–1920*. Harvard University Press.

Masika, R. (2002). *Gender, development and climate change*. Oxfam.

McCorkel, J. A., & Kristen, M. (2003). What difference does difference make? Position and privilege in the field. *Qualitative Sociology, 26*(2), 199–231.

Messer, N. M. (2003). *The role of local institutions and their interaction in disaster risk mitigation: A literature review.* Retrieved from ftp://ftp.fao.org/docrep/fao/006/ad710e/ad710e00.pdf on July 2015.

Mies, M., & Shiva, V. (1993). *Ecofeminism*. Kali for women and Zed Books.

Miescher, S. F. (2005). *Making men in Ghana*. Indiana University Press.

Milazzo, A., & Walle, D. (2017). Women left behind? Poverty and Headship in Africa. *Demography, 54*(3), 1119–1145.

Miller, B. A. (2001). *Empowering women to achieve food security: Rights to livestock*. International Food Policy Research Institute.

Ministry of Food and Disaster Management [MoFDM]. (2005). *Corporate Plan 2005–2009*. MoFDM.

Mir, M., & Bala, S. K. (2014). NGO accountability in Bangladesh: Two contrasting cases. *VOLUNTAS: International Journal of Voluntary and Nonprofit Organizations, 26*(5), 1831–1851.

Mitra, A. (2015). Severe cyclonic storm, Aila: A preliminary report. *Indian Journal of Geo-Marine Sciences, 44*, 1–26.

Mitchell, R. E., & Hodson, C. A. (1986). Coping and social support among battered women: An ecological perspective. In S. E. Hobfoll (Ed.), *Stress, social support, and women* (pp. 153–168). Hemisphere Publications.

Mitchell, T., Tanner, T., & Lussier, K. (2007). *We know what we need: South Asian women speak out on climate change adaptation*. University of Sussex.

Moreno, J., & Shaw, D. (2018). Women's empowerment following disaster: A longitudinal study of social change. *Natural Hazards, 92*, 205–224.

Morrow, B. H. (1997). Stretching the bond: The families of Andrew. In W. G. Peacock, B. H. Morrow, & H. Gladwin (Eds.), *Hurricane Andrew: Ethnicity, gender and the sociology of disasters* (pp. 141–170). Routledge.

Morrow, B. H., & Enarson, E. (1996). Hurricane Andrew through women's eyes: Issues and recommendations. *International Journal of Mass Emergencies and Disasters, 14*(1), 1–22.

Moser, C. O. N. (2014a). *Gender planning and development: Revisiting, deconstructing and reflecting*. University College London.

Moser, C. O. N. (2014b). *Gender planning and development: Theory, practice and training*. Routledge.

Mubarak, A. M., & Akram, A. A. (2016). *Seasonal migration to increase incomes of poor households in Bangladesh*. Copenhagen Consensus Centre.

Munmun, R., Sarker, A., Hoque, J., & Kabir, H. K. (2015). Women's participation in agricultural activities at forest land areas of Bangladesh. *Journal of Social Sciences, 11*(1), 30–38.

Nahar, N., Blomstedt, Y., Wu, B., Kandarina, I., Trisnantoro, L., & Kinsman, J. (2014). Increasing the provision of mental health care for vulnerable, disaster-affected people in Bangladesh. *BMC Public Health, 14*, 708–721.

Nakagawa, Y., & Shaw, R. (2005). Social capital: A missing link to discovery. *International Journal of Mass Emergencies and Disasters, 22*(1), 5–34.

Naples, A. N. (2003). *Feminism and method: Ethnography, discourse analysis and activist research*. Routledge.

Nasreen, M. (1995). *Coping with floods: The experience of rural women in Bangladesh* (Unpublished doctoral dissertation). Massey University, New Zealand.

Nasreen, M. (2003). *Gender and sustainable development in Bangladesh: Myths and realities*. Paper presented at the workshop on 'Environment and Development', Hanoi, Vietnam.

Nasreen, M. (2008). *Violence against women during disaster and post disaster situations in Bangladesh*. Action Aid International and Action Aid Bangladesh.

Nasreen, M. (2010). Rethinking disaster management: Violence against women during floods in Bangladesh. In S. Dasgupta, S. Ismail, & P. Sarathi De (Eds.), *Women's encounter with disaster* (pp. 232–244). Frontpage.

Nasreen, M. (2011). *Stepping towards equality: Social inclusion and gender*. Presented at a seminar 'Bangladesh Celebrating 40 Years of Independence', Dhaka, INGO Forum.

Nasreen, M. (2012). *Women and girls vulnerable or resilient?* Institute of Disaster Management and Vulnerability Studies, University of Dhaka.

Naved, R. T. (2011). *A rapid assessment of gender in agriculture of Bangladesh*. Centre for Diarrheal Disease Research, Bangladesh (ICDDR, B).

Nawaz, F. (2009). *Critical factors of women entrepreneurship development in rural Bangladesh* (Bangladesh Development Research Working Paper Series 5), 1–16.

Nazneen, S., & Sultan, M. (2009). Struggling for survival and autonomy: Impact of NGO-ization on women's organizations in Bangladesh. *Development, 52*(2), 193–199.

Nazneen, S., & Sultan, M. (2012). Contemporary feminist politics in Bangladesh: Taking the bull by the horns. In S. Roy (Ed.), *New South Asian feminism: Paradoxes and possibilities* (pp. 87–107). Zed Books.

Nazneen, S., Hossain, N., & Sultana, M. (2011). *National discourses on women's empowerment in Bangladesh: Continuities and change* (IDS Working Paper, 368).

Ndaruzaniye, V. (2013). *The impact of climate change on women in the African societies*. Global Water Institute.

Neelormi, S. (2010). Addressing gender issues in adaptation. In A. U. Ahmed (Ed.), *Reducing vulnerability to climate change: The pioneering example of community based adaptation* (pp. 111–127). Center for Global Change and CARE, Bangladesh.

Nelson, V., Meadows, K., Cannon, T., Morton, J., & Martin, A. (2002). Uncertain prediction, invisible impacts and the need to mainstream gender in climate change adaptations. *Gender and Development, 10*(2), 51–59.

Neuman, W. L. (1997). *Social research methods: Qualitative and quantitative approaches.* Allyn and Bacon.

Neuman, W. L. (2006). *Social research methods: Qualitative and quantitative approaches.* Pearson.

Nguyen, M. C., &Wodon, Q. (2015). Temporary and permanent migration. In A. O'Donnell & Q. Wodon (Eds.), *Climate change adaptation and social resilience in the Sundarbans* (pp. 123–150). Taylor and Francis.

Nibi, S. H. (2012). *A feminist political ecology of large-scale agro fuel production in Northern Ghana: A case study of Kpachaa.* International Institute of Social Studies.

Nightingale, A. (2006). The nature of gender: Work, gender, and environment. *Environment and Planning: Society and Space, 24*, 165–185.

Ninno, C., del., Dorosh, P. A., Smith, L. C., & Roy, D. K. (2001). *The 1998 floods in Bangladesh disaster impacts, household coping strategies, and response.* International Food Policy Research Institute.

Nonoguchi, A. (2012). *Gender and climate in Nepal* (Unpublished doctoral dissertation). The Graduate School College of Agricultural Science: The Pennsylvania State University.

Noor, J. (2016). *Women leadership in emergency.* Presented in training programme on 'Women's Leadership in Emergency.' Oxfam Bangladesh.

Norris, F. H., Friedman, M. J., Watson, P. J., Byrne, C. M., Diaz, E., & Kaniasty, K. (2002). Disaster victims speak: Part I. An empirical review of the empirical literature. *Psychiatry, 65*(3), 207–239.

Norris, F. H., Stevens, S. P., Pfefferbaum, B., Wyche, K. F., & Pfefferbaum, R. L. (2008). Community resilience as a metaphor, theory, set of capacities, and strategy for disaster readiness. *Journal of Community Psychology, 41*(1–2), 127–150.

O'Donnell, A., & Wodon, Q. (2015). *Climate change adaptation and social resilience in the Sundarbans.* Taylor and Francis.

O'Dwyer, B., & Unerman, J. (2008). The paradox of greater NGO accountability: A case study of amnesty Ireland. *Accounting, Organizations, and Society, 33*, 801–824.

Odendaal, W. (2001). The men against violence against women movement in Namibia. *Development, 44*, 90–93.

Oxfam. (1995). *The Oxfam handbook of relief and development.* Oxfam.

Oxfam. (2005). The tsunami's Impact on Women. *Oxfam Briefing Note.* Oxfam.

Oxfam. (2011). *Women leadership in disaster management.* Oxfam.

Pagiola, S. (1995). *Environmental and natural resource degradation in intensive agriculture in Bangladesh* (Environmental Economics Series, Paper No.15). Environmentally Sustainable Development: World Bank.

Pakulski, J. (2004). *Globalizing inequalities.* Allen and Unwin.

Panday, P. (2016).*Women's empowerment in South Asia: NGO interventions and agency building in Bangladesh.* Routledge.

Paolisso, M., & Amanda, R. (2002). The significance of the gender division of labor in assessing disaster impacts: A case study of hurricane Mitch and Hillside farmers in Honduras. *International Journal of Mass Emergencies and Disasters, 20*(2), 171–195.

Parveen, S. (2007). Gender awareness of rural women in Bangladesh. *Journal of International Women's Studies, 9*(1), 253–269.

Parvin, G. A., Ahsan, S. M. R., & Chowdhury, M. R. (2005). Women empowerment performance of income generating activities: A case study in Dumuria Thana, Bangladesh. *The Journal of Geo-Environment, 4*, 47–62.

Paul, B. K. (2009). Why relatively fewer people died? The case of Bangladesh's cyclone *Sidr. Natural Hazards, 50*(2), 289–304.

Peacock, D. (2003). Building a legacy on social justice and activism. *Men and Masculinities, 5*, 325–328.

Peacock, W. G., Morrow, B. H., & Gladwin, H. (1997). *Hurricane Andrew: Ethnicity, gender, and the sociology of disasters.* Routledge.

Peek, L. (2008). Children and disasters: Understanding vulnerability, developing capacities, and promoting resilience—An introduction. *Child Youth Environment, 18*(1), 1–29.

Peek, L., Morrissey, B., & Marlatt, H. (2011). Disaster hits home: A model of displaced family adjustment after hurricane Katrina. *Journal of Family Issues, 32*(10), 1371–1396.

Peek, L., & Stough, L. M. (2010). Children with disabilities in the context of disaster: A social vulnerability perspective. *Child Development, 81*(4), 1260–1270.

Penrose, A., & Takaki, M. (2006). Children's rights in emergencies and disasters. *The Lancet, 367*, 698–699.

Phellas, C. N. (2006). Review: Keith F. Punch (2005). Introduction to social research: Quantitative & qualitative approaches. *Forum: Qualitative Social Research, 7*(2).

Plummer, K. (1983). *Documents of life: An introduction to the problems and literature of a humanistic method.* Allen and Unwin.

Puigvert, L. (2003). Dialogic feminism: "Other women's" contributions to the social transformation of gender relations. In E. Beck-Gernsheim, J. Butler, & L. Puigvert (Eds.), *Women and social transformation* (pp. 29–60). Peter Lang.

Punch, K. F. (2005). *Introduction to social research: Quantitative and qualitative approaches.* Sage.

Punch, K. F. (2006). *Developing effective research proposals.* Sage.

Rabbani, M. G. (2009). Climate force migration: A massive threat to coastal people in Bangladesh. In *Clime Asia: Climate Action Network-South Asia (CANSA) Newsletter* (p. 4). Bangladesh Centre for Advanced Studies.

Rahman, K. M. A. (2014). Globalization and cultural transformation: The case of Bangladesh. *Asian Culture and History, 6*(2), 1–10.

Rahman, M. H., & Naoroze, K. (2007). Women empowerment through participation in aquaculture: Experience of a large-scale technology demonstration project in Bangladesh. *Journal of Social Science, 3*(4), 164–171.

Rahm, E., & Do, H. H. (2000). *Data cleaning: Problems and current approaches.* Retrieved from http://dbs.uni-leipzig.de/file/TBDE2000.pdf on 2 April 2015.

Rashid, M. U., & Islam, M. M. (2011). Women's participation in family decision making in DumkiUpazila of Patuakhali district. *The Agriculturist, 1*(2), 137–142.

Rashid, S. F., & Michaud, S. (2000). Female adolescents and their sexuality: Notions of honour, shame, purity and pollution during the floods. *Disasters, 24*(1), 54–70.

Reddy, S. (2016, March 9). Increasing women's access to skills and jobs. *The Daily Star.* Retrieved from http://www.thedailystar.net/op-ed/increasing-womens-access-skills-and-jobs-788224 on January 22, 2017.

Reed, M. G., & Mitchell, B. (2003). *Gendering environmental geography.* Retrieved from onlinelinelibrary.wiley.com on October 10, 2019.

Reyes, R. R. (2002). Gendering response to El Nino in rural Peru. *Gender and Development, 10*(2), 60–69.

Rice, A. E., & Ritchie, C. (1995). Relationships between international non-governmental organizations and the United Nations. *Transnational Associations, 47*(5), 254–265.

Richter, R., & Flower, T. (2010). Gender aware disaster care: Issues and interventions in supplies, services, triage, and treatment. *International Journal of Mass Emergencies and Disasters, 28*(2), 207–225.

Ripley, A. (2008). *The unthinkable: Who survives when disaster strikes and why?* Crown.

Rocheleau, D., Thomas-Slayter, B., & Wangari, E. (1996). *Feminist political ecology: Global issues and local experiences.* Routledge.

Rosa, K. (2005). *Voices from the wave.* Fast Print.

Roy, D. R. (2011). Vulnerability and population displacements due to climate induced disasters in coastal Bangladesh. In M. Leighton, X. Shen, & K. Warner (Eds.), *Climate change and migration: rethinking policies for adaptation and disaster risk reduction* (pp. 24–33). Studies of the University: Research, Counsel, Education.

Roy, K. (2004). *Water logging in the South-West coast: Reducing vulnerability to climate change.* Climate Information Cell.

Roy, M., & Venema, D. (2002). Reducing risk and vulnerability to climate change in India: The capabilities approach. *Gender and Development, 10*(2), 78–83.

Rozario, S. (1992). *Purity and communal boundaries: Women and social changes in Bangladeshi village*. Allen and Unwin.

Ruether, R. R. (1993). Ecofeminism: Symbolic and social connections of the oppression of women and the domination of nature. In C. J. Adams (Ed.), *Ecofeminism and the sacred* (pp. 13–23). The Continuum.

Rupantor. (2013). *Political empowerment of women at grass root level: Rupantor model*. Rupantor.

Saguna, B. (2006). *Empowerment of rural women through self help groups*. Discovery.

Saha, C. K. (2015). Dynamics of disaster-induced risk in southwestern coastal Bangladesh: An analysis on tropical Cyclone *Aila* 2009. *Natural Hazards, 75*(1), 727–754.

Saifullah, M., & Ahammad, S. M. (2008). Accountability of NGO operations in Bangladesh. *Journal of Science and Technology., 6*, 28–36.

Sayeed, A. T. (2010). Victims of earthquake and patriarchy: The 2005 Pakistan earthquake. In E. Enarson & P. G. Dhar Chakrabarti (Eds.), *Women, gender and disaster: Global issues and initiatives* (pp. 142–151). Sage.

Schendel, W. V. (2009). *A history of Bangladesh*. Cambridge University Press.

Sen, A. (1981a). *Poverty and famines: An essay on entitlement and deprivation*. Clarendon Press.

Sen, A. (1981b). Ingredients of famine analysis: Availability and entitlements. *The Quarterly Journal of Economics, 96*(3), 433–464.

Sen, A. (1984). *Resources, values and development*. Blackwell.

Sen, A. (1985). Well being, agency and freedom. *Journal of Philosophy, 132*(4), 169–221.

Sen, A. (1988). Family and food: Sex bias in poverty. In T. Srinivasav & P. Bardhan (Eds.), *Rural poverty in South Asia* (pp. 453–472). Columbia University Press.

Sen, A. (2000). *Development as freedom*. Alfred Knop.

Shamsuddoha, M., Islam, M., Haque, M. A., Rahman., M. F., Roberts, E., Hasemann, A., & Roddick, S. (2013). *Local perspective on loss and damage in the context of extreme events: Insights from cyclone-affected communities in coastal Bangladesh*. Centre for Participatory Research and Development.

Shamsuddoha, M., & Chowdhury, R. K. (2007). *Climate change impact and disaster vulnerabilities in the areas of Bangladesh*. COAST Trust.

Shaw, R. (2006). *Indian Ocean tsunami and aftermath: Need for environment disaster synergy in the reconstruction process*. Retrieved from http://www.emaraldinsight.com/0965-3562.htm on 2 June 2015.

Shaw, J., Applegate, B., Tanner, S., Perez, D., Rothe, E., Bowen, A. C., & Lahey, B. (1995). Psychological effects of hurricane Andrew on an elementary school population. *Journal of the American Academy of Child and Adolescent Psychiatry, 34*(9), 1185–1192.

Sheheli, S. (2012). *Improving livelihood of rural women through income generating activities in Bangladesh* (Unpublished doctoral dissertation). Berlin: University of Berlin.

Shiva, V. (1988). Staying alive. In *Women, ecology and development*. Zed Books.

Shreve, C., Davis, B., & Fordham, M. (2017). Gender and disaster. In D. Paton & D. Johnston (Eds.), *Disaster resilience: An integrated approach* (pp. 175–193). Charles Thomas.

Singh, K. (2014). Application of Pressure and Release (PAR) model for assessing vulnerability to industrial hazards in district Bathinda (Punjab, India). *International Journal of Management and Social Sciences Research, 3*(5), 25–29.

Smit, B., & Wandel, J. (2006). Adaptation, adaptive capacity and vulnerability. *Global Environmental Change, 16*, 282–292.

Sobhan, R. (2012). *Challenging the injustice of poverty*. Sage.

Sorensen, J. H., & White, G. F. (1980). Natural hazards: A cross cultural perspective. In I. Attman, A. Rapoport, & J. F. Wohlwill (Eds.), *Human behavior and environment: Advances in theory and research*. Plenum Press.

Staskiewics, J., & Khan, S. I. (2013, April). *Bangladesh's enhancing resilience programme*. Paper presented in the conference 'Hunger nutrition climate justice 2013: A New Dialogue: Putting people at the heart of the global development. Dublin, Ireland.

Strauss, A., & Corbin, J. (1990). *Basics of qualitative research: Grounded theory procedures and techniques*. Sage.

Strauss, A., & Corbin, J. (1998). *Basics of qualitative research: Techniques and procedures for developing grounded theory*. Thousand Oaks.

Sultana, A. M. (2010). Patriarchy and women's gender ideology: A socio-cultural perspective. *Journal of Social Sciences, 6*(1), 123–126.

Sultana, A. M. (2011). Factors effect on women autonomy and decision making power within the household in rural communities. *Journal of Applied Sciences Research, 7*(1), 18–22.

Sundberg, J. (2017). Feminist political ecology. In D. Richardson, N. Castree, M. F. Goodchild, A. Kobayashi., W. Liu, & R. A. Marston (Eds.), *International encyclopedia of geography: People, the earth, environment and technology* (pp. 1–10). Wiley.

Szapuova, M. (2006). Mills liberal feminism: Its legacy and current criticism. *Prolegomena, 5*(2), 179–191.

Tanner, T., & Allouche, J. (2011). Towards a new political economy of climate change and development. *IDS Bulletin, 42*(3), 1–14.

Tashakkori, A., & Teddlie, C. (2003). *Handbook of mixed methods in social & behavioral research*. Sage.

Terry, G. (2009). No climate justice without gender justice: An overview of the issues. *Gender and Development, 17*(1), 5–18.

Thieme, S., & Siegmann, K. (2010). Coping on women's backs: Social capital- vulnerability links through a gender lens. *Current Sociology, 58*(5), 719–737.

TIB [Transparency International Bangladesh]. (2012). *Corruption in service sectors: National household survey 2012*. TIB.

TIB [Transparency International Bangladesh]. (2015). *Corruption in service sectors: National household survey 2015*. TIB.

TIB [Transparency International Bangladesh]. (2018). *NGOs in Bangladesh funded by foreign donations: Governance challenges and way forward*. TIB.

Tierney, K. (2006, March). *Hurricane Katrina: Catastrophic impacts and alarming lessons*. In Conference paper series prepared for Berkeley symposium on real estate, catastrophic risk, and public policy.

Tobin-Gurley, J., Peek, L., & Loomis, J. (2010). Displaced single mothers in the aftermath of Hurricane Katrina: Resource needs and resource acquisition. *International Journal of Mass Emergencies Disasters, 28*(2), 170–206.

Turner, B. L., Kasperson, R. E., Matsone, P. A., McCarthy, J. J., Corell, R. W., Christensene, L., … Schiller, A. (2003). A framework for vulnerability analysis in sustainability science. *The Proceedings of National Academy of Sciences, 100*(4), 8074–8079.

UNDP [United Nations Development Programme]. (2018). Enhancing adaptive capacities of coastal communities, especially women, to cope with climate change induced salinity. *Gender Assessment.*

UNICEF [United Nations International Children's Emergency Fund]. (2015). *Committing to child survival: A promise renewed*. UNICEF.

UnnayanOnneshan. (2009). *Initial Assessment Report: With Focus on Khulna District*. Dhaka: UnnayanOnneshan.

UnnayanOnneshan. (2010). *Cyclone Aila: One year on natural disaster to human sufferings*. UnnayanOnneshan.

United Nations Economic and Social Commission for Asia and the Pacific [UN ESCAP]. (2010). *Asia Pacific Disaster Report 2010*. UN ESCAP.

UN WFP [United Nations World Food Programme]. (2006). *Rural Bangladesh: Socio-economic profiles of WFP operational areas & beneficiaries*. TANGO International.

UN Women. (2016). *Time to act on gender, climate change and disaster risk reduction. UN Women: Asia and the Pacific*. Retrieved from http://asiapacific.unwomen.org/en/digital-library/publications/2016/11/time-to-act on May 26, 2019.

United Nations Division for the Advancement of Women (UN DAW). (2004). *Making risky environments safer. Women 2000 and beyond*. Department of Economic and Social Affairs: UN DAW.
United Nations International Strategy for Disaster Reduction [UN ISDR]. (2004). *Living with risk: A global review of disaster reduction initiatives*. UN Publications, United Nations.
United Nations International Strategy for Disaster Reduction [UN ISDR]. (2012). *Disaster and social vulnerabilities in Asia and the Pacific*. Retrieved from www.duryognivaran.org on March 31, 2015.
Vance, L. (1997). Ecofeminism and wilderness. *National Women's Studies Association Journal, 9*(3), 60–76.
Varga, C. A. (2001). The forgotten fifty percent: A review of sexual and reproductive health research and programs focused on boys and young men in Sub-Saharan Africa. *African Journal of Reproductive Health, 5*, 175–195.
Vinas, C. S. (1998). Women's disaster vulnerability and response to the Colima earthquake. In E. Enarson & B. H. Morrow (Eds.), *The gendered terrain of disaster: Through women's eye* (pp. 161–172). Praeger Publishers.
Visvanathan, N., Duggan, L., Nisonoff, L., & Wiegersma, N. (Eds.). (1997). *The women, gender, and development reader*. Zed Books.
Wallace, T., Bornstein, L., & Chapman, J. (2006). *The aid chain: Coercion and commitment in development NGOs*. ITDG.
Wamsler, C., Brink, E., & Rentala, O. (2012). Climate change, adaptation, and formal education: The role of schooling for increasing societies' adaptive capacities in El Salvador and Brazil. *Ecology and Society, 17*(2), 1–19.
Watkins, S. C., Swidler, S., & Hannan, T. (2012). Outsourcing social transformation: Development NGOs as organizations. *Annual Review of Sociology, 38*, 285–315.
Warren, K. J. (1996). The power and the promise of ecological feminism. In K. J. Warren (Ed.), *Ecological feminist philosophies* (pp. 19–41). Indiana University Press.
Warren, K. J. (1997). Taking empirical data seriously: An economist philosophical perspectives. In K. J. Warren (Ed.), *Ecofeminism: Women, culture, nature* (pp. 3–20). Indiana University Press.
WEDO [Women Economic Development Outreach]. (2008). *Gender, climate change and human security lessons from Bangladesh, Ghana and Senegal*. Retrieved from www.gdnonline.org on November 24, 2016.
West, C., & Zimmerman, D. H. (1987). Doing gender. *Gender and Society, 1*(2), 125–151.
Wheeler, E. F., & Abdullah, M. (1988a). Food allocation within the family. In I. de Garine & G. A. Harrison (Eds.), *Coping with uncertainty in the food supply* (pp. 335–353). Clarendon Press.
Wickberg, S. (2012). *Overview of corruption and anti-corruption in Bangladesh*. Transparency International.
William, A., & Taylor, J. (2013). Resolving accountability ambiguity in nonprofit organizations. *International Journal of Voluntary and Nonprofit Organizations, 24*, 559–580.
Wilson, J., Phillips, B. D., & Neal, D. M. (1998). Domestic violence after disaster. In E. Enarson & B. H. Morrow (Eds.), *The gendered terrain of disaster: Through women's eye* (pp. 81–94). Praeger Publishers.
Wisner, B., Blaikie, P., Cannon, T., & Davis, I. (2004). *At risk: Natural hazards, people's vulnerability and disasters*. Routledge.
Wheeler, E. F., & Abdullah, M. (1988b). Food allocation within family: Response to fluctuating food supply and food needs. In I. D. E. Garine & G. A. Harrison (Eds.), *Coping with uncertainty of food supply*. Clarendon Press.
WHO [World Health Organization]. (2012). *Mainstreaming gender in health adaptation to climate change programmes*. Switzerland.
Wollstonecraft, M. (1992). *A vindication of the rights of woman*. Everyman's Library.
World Bank. (2006a). *Bangladesh country assistance strategy 2006a–2009*. World Bank.
World Bank. (2006b). *Bangladesh country environmental analysis*. Bangladesh Development Series, 12. World Bank Office.

World Bank. (2008). *Poverty assessment for Bangladesh: Creating opportunities and bridging the East-West divide.* The World Bank.

World Bank. (2009). *Whispers to voices: Gender and social transformation in Bangladesh.* Bangladesh Development Series, 22.

Yohe, G., Strzepek, K., Pau, T., & Yohe, C. (2003). Assessing vulnerability in the context of changing socioeconomic conditions: A study of Egypt. In J. B. Smith, R. J. T. Klein, & S. Huq (Eds.), *Climate change, adaptive capacity and development*. Imperial College Press.

Zahran, S., Peek, L., Snodgrass, J. G., Weiler, S., & Hempel, L. (2013). Abnormal labor outcomes as a function of maternal exposure to a catastrophic hurricane event during pregnancy. *Natural Hazards, 66*(1), 61–76.

Zahran, S., Shelley, T. C., Peek, L., & Brody, S. D. (2009). Natural disasters and social order: Modeling crime outcomes in Florida. *International Journal of Mass Emergencies Disasters, 27*(1), 26–52.

Zaman, H. (2009). *Assessing the poverty and vulnerability impact of micro credit in Bangladesh: A case study of BRAC.* The World Bank.

Zaman, M. Q. (1999). Vulnerability, disaster, and survival in Bangladesh: Three case studies. In A. Oliver-Smith & S. M. Hoffman (Eds.), *The angry earth: Disaster in anthropological perspective* (pp. 192–212). Routledge.

The manufacturer's authorised representative in the EU is Springer Nature Customer Service Centre GmbH, Europaplatz 3, 69115 Heidelberg, Germany. If you have any concerns regarding our products, please contact ProductSafety@springernature.com

Printed and bound by CPI Group (UK) Ltd, Croydon, CR0 4YY
02/10/2026
02227130-0004